Environment and Tribes in India

Resource conflicts and adaptations

ABOUT THE EDITOR

Vineetha Menon, Former Head of the Department of Anthropology and Dean of Humanities, is Professor of Anthropology and co-ordinator of the UGC Special Assistance Programme on Tribes of Kerala: Development and Change in Kannur University, Kerala, India. After post-graduate degrees in Sociology and in Anthropology from Jawaharlal Nehru University and the University of British Columbia respectively, she obtained Ph.D in Cultural Anthropology from York University (Canada) where she was a Commonwealth Scholar (1990-95). Her research interests are both wide and focused. Interested in both theory and practice, she has worked on a variety of issues ranging from those related to anthropological theory and methods to interdisciplinary and multidisciplinary issues of Inclusive and Sustainable Development, Governance and Decentralisation, Natural Resource Management, Gender Justice, Marginalised Populations, Social Movements, Anthropology of Health and Healing, Utilisation of Knowledge, etc.

Among her publications are the volumes: Alleviating Poverty—Case Studies of Local-level Linkages and Processes in the Developing World (co-edited with PR Gopinathan Nair and KN Nair); Confronting Violence Against Women—Engendering Kerala's Development Experience (edited with K.N. Nair); Water Insecurity, Institutions and Livelihood Dynamics: A Study in Plachimada, Kerala, (with Antonyto Paul and K.N. Nair); and, Social Change in Kerala-- Insights from Micro-level Studies (edited with K.N. Nair) and many articles in peer reviewed national and international journals.

Environment and Tribes in India

Resource conflicts and adaptations

Edited by

Vineetha Menon

Indira Gandhi Rashtriya Manav Sangrahalaya, Bhopal

CONCEPT PUBLISHING COMPANY PVT. LTD., NEW DELHI-110059

ISBN-13: 978-81-8069-872-9

First Published 2012

Published and Printed by

Indira Gandhi Rashtriya Manav Sangrahalaya
(National Museum of Mankind)
Shamla Hills, Bhopal - 462013 (India)

and

Concept Publishing Company Pvt. Ltd.
A/15-16, Commercial Block, Mohan Garden
New Delhi-110059 (India)
Editorial Office: H-13, Bali Nagar, New Delhi-110 015, India.

Cataloging in Publication Data--*Courtesy:* D.K. Agencies (P) Ltd. <docinfo@dkagencies.com>

Environment and tribes in India : resource conflicts and adaptations / edited by Vineetha Menon.

p. cm.
Includes bibliographical references and index.
ISBN 9788180698729

1. Adivasis—India—Congresses. 2. Adivasis—India—Kerala—Congresses. 3. Human ecology—India—Congresses. 4. Human ecology—India—Kerala Congresses. 5. Resource-based communities—India—Congresses. 6. Resource based communities—India—Kerala—Congresses. 7. Natural resources—Government policy—India—Congresses. 8. Natural resources—Government policy—India Kerala Congresses. I. Menon, Vineetha. II. Indira Gandhi Rashtòriya Manaba Samògrahalaya.

DDC 305.800954 23

Foreword

Anthropological discourse on the issue of human perceptions of nature has been much debated in recent times without any satisfactory final answer as to how the natives think about their environment. However, rich ethnographic studies aimed at underscoring human perceptions of nature have finally put to an end the controversial Western notion of ontological dualism of 'nature' and 'culture' in favour of a monistic paradigm. With the decline of Cartesian dualism and due to the influence of phenomenological perspective, it is now postulated that the perception of nature is primarily socially and culturally constructed rather than subjected to a predisposed *a priori* analytical framework. This theoretical position ostensibly leads to two mutually opposed approaches, a relativistic-deconstructionist approach and a universalistic-reductionist approach to understand the human perception of nature.

The hard relativistic stand has a long and protracted legacy, and Edmund Leach in 1964 had suggested: "I postulate that the physical and social environment of a young child is perceived as a continuum. It does not contain any intrinsically separate 'things'. The child, in due course, is taught to impose upon this environment a kind of discriminating grid which serves to distinguish the world as being composed of a large number of separate things, each labeled with a name. The world is a representation of our language categories, not *vice versa*". Much later, Marilyn Strathern endorses this position by saying, "...there is no such thing as nature or culture. Each is a highly relativized concept whose ultimate signification must be derived from its place within a specific metaphysics. No single meaning can in fact be given to nature or culture in Western thought; there is no consistent dichotomy, only a matrix of contrasts". Many ethnoecological researches reinforce this stand with many different perceptions of culturally significant materials in different cultures all over the world.

The notion of universal human cognitive predisposition goes back to John Locke though, it was given a fresh lease of life in the writings of Noam Chomsky. Chomsky's argument is based on an innate programmed human capacity for language, which gains further support from the advances in neuro-physiology that language faculty is localized in the human brain to make it a speaking animal. Even the findings in development psychology, particularly in the writings of Richard Schweder, give credence to the fact that there are certain kinds of universal cognitive dispositions that account for cross-cultural recurrence of some mental representations, which humans might have got through the process of natural selection. Any attempt, therefore, according to Philip Descola, to understand human perception of nature using the universalistic approach finally crumbles to a model of 'regional metasystem' whose application beyond a specific region remains extremely polemical. This is precisely the strength of empirical studies that anthropology professes and anthropologists invariably undertake. Professor Vineetha Menon has done a commendable job of neatly knitting together seventeen pieces of empirical works on the relationship between ecology, indigeneity, resource conflicts and adaptations in India, besides her introduction, in this volume.

Professor Menon in her introduction aptly confesses that "indigeneity" is a much contested concept. But since it has been in use in transnational discourse overriding national preferences, I do not find any reason for being so apologetic about it. Perhaps I would consider indigeneity more as a value or a worldview or an idea rather than an ethnic group or community. In Indian parlance, it proximates to the ideas or perceptions of the Adivasis about their ecology, conflicts arising out of resource depletion or changing land use in their habitats, and eventual process of adaptation to the changing ecology and scarce resources.

It is too well known that Adivasis in India are multi-ethnic, multi-cultural and multi-linguistic communities. But one important characteristic most of the Adivasis share is that wherever they live, they are at the bottom of economic and social hierarchies, and are the poorest, most marginalized, oppressed, and deprived people in this country. This has once led Prof. L.K.Mahapatra to forcefully conclude that the prominence of indigenous peoples' concerns stems from the realization that they have meagerly benefited from many development projects, while the mainstreamed societies have prospered at their expense, pushing them deeper into the poverty trap. It is an irony that poverty and economic backwardness are invariably tagged up with the Adivasis, although they inhabit the most resource-rich regions in our country and elsewhere. They have physical, spiritual and psychological attachment to these resources, be it land, forest or water. Mindless implementation of mega development and ecotourism projects in Adivasi inhabited regions ignoring their perception and participation has culminated in disasters for these people.

Most of the chapters included in this volume largely revolve around the above theme directly or discreetly. They echo the perceptions and articulate the expectations of the Adivasis from Kerala, Assam, Puducherry, West Bengal, Odisha and other parts of the country with regard to ecology, resources and adaptive strategies. I am sure these empirical essays will be of immense interest to scholars in anthropology and sociology, tribal studies, resource management, policy makers, and the NGOs working in the Adivasi inhabited areas. On behalf of Indira Gandhi Rashtriya Manav Sangrahalaya, Bhopal, and on my personal behalf, I congratulate Professor Menon for editing such a useful volume with all pain and care.

Bhopal
18 February 2012

Prof. Kamal K. Misra
Director, IGRMS

Acknowledgements

This volume results from two seminars organized under the University Grants Commission's Special Assistance Programme in the Department of Anthropology, Kannur University on the theme: Ecology, Indigeneity, Resource Conflicts and Adaptations: India and Kerala. Indian Council of Social Science Research, New Delhi and Kannur University provided additional funding. First and foremost, my gratitude to Indira Gandhi Rashtriya Manav Sangrahalaya (IGRMS), Bhopal and its Director Prof. KK. Misra for agreeing to publish this volume is placed on record. As University Grants Commission's nominee to the Advisory Board of the Special Assistance Programme in the Department, Prof. Misra has closely associated with our academic activities and has extended support. My sincere thanks to him. I also acknowledge my gratitude to UGC, ICSSR and Kannur University for the financial support that allowed the coming together of scholars from various parts of India. Needless to say, the deliberations benefited from such diversity of experience and expertise. I am also grateful to all the participants of the seminars for bringing local issues of their concern for vibrant discussions in a comparative framework. While the first seminar related to situations all over India, the second one pertained specifically to indigenous populations in Kerala. Although the concept note of the seminars mentioned focus on indigeneity, we knew well that it is much contested and that in any issue relating to human ecology or human conflicts or human identity, strict compartmentalization is never possible and that these categories are overlapping and mutually constituted. The papers also proved this to be true. I am indebted to all the authors who permitted me to include their papers in this volume and gave me the liberty to edit them. Many of them have painstakingly reworked on their papers reflecting on questions that emerged in the course of discussions. I thank them. Susan Visvanathan deserves special mention for allowing me to publish her keynote address to the seminar.

Prof. Michael Tharakan has provided much motivation in organizing the seminar and the follow-up work of documentation, not only in his official capacity as the Vice-Chancellor of Kannur University, but also as a scholar historian who constantly urges academic exchanges across regional and disciplinary bounds. His encouragement has not been confined to being Chairperson of the Advisory Committee of UGC-SAP in the Department of Anthropology, but also as a committed academic who keeps alive his interest in serious research and scholarship despite the burden of administrative duties. I place on record my appreciation for his constant encouragement to our academic efforts.

Seetha Kakkoth and Shibu, P. have joined me in the laborious manuscript correction with proof reading and in communicating with several authors. Without the definite source of their support, I would not have ventured into this task. I cannot thank them enough. My colleagues Dr S. Gregory, Dr. B. Bindu, Dr. M. Sini and Dr. M.S. Mahendrakumar, as well as several students of the department of Anthropology in Kannur University have contributed in various ways in the organizing of the seminars in which these papers were presented. I thank them all.

Vineetha Menon

Contents

List of Contributors

A.M., Prasad, Research Assistant, Kerala Institute for Research, Training and Development Studies of Scheduled Castes & Scheduled Tribes (KIRTADS), Kozhikode. Kerala, prasadpoothadi@gmail.com

Amalmani Sarma, C/o A.K. Deka Lachit Nagar, ByeLane-8, Sub Bye Lane-6 (Right), House No. 4, PO Ulubari, Guwahati-7, Assam, amalmani_79@yahoo.co.in

B. Francis Kulirani, Superintending Anthropologist, Anthropological Survey of India, Mysore, fkulirani@ yahoo.com

C.P., Vinod, Senior Academic Consultant, School of Interdisciplinary and Transdisciplinary Studies, IGNOU, Delhi, vinod.vel@gmail.com

K.P., Suresh, Research Assistant, Dept. of KIRTADS, Kozhikode. Kerala, sureshpanathur@gmail.com

M., Amruth, Scientist, Forestry and Human Dimension Programme Division, Kerala Forest Research Institute, Peechi, Thrissur, Kerala, amrutham@gmail.com

Mohan Kumar Bera, Research Scholar, Tata Institute of Social Sciences, Mumbai-400088, mohan.bera@gmail.com

P.M. Nanda, Reader in Anthropology, Govt. College, Phulbani, Orissa.

P.R. Sahoo, Research Scholar, Institute of Rural Reconstruction, Visva Bharati University, West Bengal, piyusa@gmail.com

Dr. P.R.G. Mathur, Secretary, Anantha Krishna Iyer Centre for Advance Studies, Palakkad, Kerala, drprgmathur@ yahoo.co.in

P.V., Mini. Research Officer (Anthropology), Dept. of KIRTADS, Kozhikode, Kerala, minipv07@gmail.com

R. Rajendra Prasad, Research Associate, Anthropological Survey of India, Mysore, rrprasadasi@yahoo.com

Dr. S. Bindu, Director, Kerala Institute for Research, Training and Development Studies of Scheduled Castes & Scheduled Tribes (KIRTADS), Kozhikode. Kerala, bindoos_s@yahoo.com

S.K., Sashikumar. Research Assistant, Kerala Institute for Research, Training and Development Studies of Scheduled Castes & Scheduled Tribes (KIRTADS), Kozhikode. Kerala, shashikasaragodu@gmail.com

Prof. Susan Visvanathan, Professor, Centre for the Study of Social Systems, Jawaharlal Nehru University, New Delhi. susanvisvanathan@gmail.com

T.V. Geetha Devi, Independent Research Scholar, Associated with Inter-university Centre for Social Science Research and Extension, M. G. University, Malloossery, P.O, Kottayam, Kerala, geeth.devi@gmail.com

Dr. V. Pragathi, Assistant Professor, Department of Anthropology, Puduchery Institute of Linguistics and Culture, Puduchery., pragathirajkumar@yahoo.co.in

V.S. Subash, Research Officer, Dept. of KIRTADS, Kozhikode. Kerala, vssubuji@rediffmail.com

1
Introduction

Vineetha Menon

The role of environment in securing subsistence and in sustainable development is crucial. The more fragile an environment is, the more are the subsistence conflicts and the challenges to sustainable development. An environment or ecosystem and thereby culture, becomes fragile on account of various factors ranging from climatic change, over-population induced by migration, changes in land use or livelihood practices, technological changes, over-exploitation, depletion and destruction of resources, etc. As the well-known anthropologist Marvin Harris had noted as early as in the 1970s, even hunting-gathering and agricultural societies lapsed into endemic warfare and gender violence when their resource bases were depleted and over-stretched. An ecological system under stress renders people vulnerable. Already vulnerable groups like the indigenous/'tribal' people, women and children, the old and the disabled especially, are the worst sufferers of such environmental stress. Migration of non-tribal people into tribal areas has been one known factor for resource conflicts in tribal areas that have pushed tribal people in groups or in smaller fragmented groups or as individual beings, to migrate to other areas seasonally or permanently, stressing out their culture, society, families and individual lives. Industries, hydro-electric or irrigation projects and other aspects of 'development' have also interfered with the lives of many tribal people all over India. Administrative interventions of State too have caused varieties of ecological stresses. Some cope better than others. However, some tribal areas in India and elsewhere have become the hotbeds of inter-ethnic conflicts within and between tribes, and between tribal people and non-tribal people. Experiences varied from adaptations to long drawn-out resistances. In some areas, even political radicalism has manifested. Violence and terror in some extreme cases, changes in notions of ethnicity in others, worsening of vulnerability in most cases, have happened in response to ecological stress and efforts at adaptations–economic, technological, cultural, religious or political, among others. Recent reports about terrorist activities in tribal habitats and protracted insurgency situations cause alarm over the worsening fate of the already exploited and vulnerable tribal people. Natural disasters on account of land-use changes or other ecological stressors are also major threats causing alarm and asking for preparedness to deal with the onslaughts.

Many anthropologists have highlighted that interventions and intrusions into indigenous populations' habitats and their lives have contributed to ecological stress as indigenous populations did not over exploit their habitat and had harmonious adaptations to environmental changes. Others have pointed out that indigeneity did not always mean peaceful co-existence with nature. In the contemporary times of increased and intensified global interactions, how do indigenous people like the *adivasis*/tribal people deal with their ecosystems, especially when States and dominant communities desire wealth from tourism promotion even in tribal habitats? How do contemporary governance structures and administrative policies impinge on the local people's access to and control over the resources in their habitat? In situations of conflicts over resources, indigeneity itself is constantly questioned as people are displaced, over and over again, over territories, or they migrate in search of better lives. It is this context of global-local, inter-local and intra-local interlinkages, conflicts and co-operation around the dialectic of ecological and commercial concerns that frames the bringing together of the chapters in this volume.

Susan Visvanathan, in her discussion of regenerative forest policy and its success in the case of Tiruvannamalai, points out that when it comes to the question of forests, tribe and caste communities may be severally divided or have shared interests. In the twentyfirst century, most of the tribal culture questions are entangled around the problems of nation-state politics, arguments for integration and assimilation, quests for nation and sovereignty, and the commercialization of their territories and products through tourism and the forces of globalization, including accessibility to territories by police, administrators, hoteliers and political parties, Susan encapsulates. There are multiple layers of meanings in understanding human-ecological relationships. Conflicts over resources are escalating; so are efforts at conservation and regeneration of nature. There is a curious mix of commercial and spiritual in contested spaces. All these are critical in understanding indigeneity, ecology and conflicts over resources. Theoretical discussions viewing ecology as common property resource, centre around ownership, management, transaction costs, and other such largely economic concerns. Use efficiency and cost reduction or durability of the institutional framework or the rules of the game rather than concerns for justice or cultural values have largely dominated the discussions. This volume, however, brings together the case studies of local level empirical issues touching the lives of mostly marginalized indigenous people. Like natural resources, indigeneity is also a heavily contested domain. Conflicts at the cross roads of ecology and indigeneity take many forms and expressions. The complicity involved in any negotiation or solution is brought out by the chapters cumulatively. The chapter are arranged in a thematic order, starting with an overarching frame involving State, market, pilgrims, tourists, and ecological, commercial, spiritual and cultural concerns, and ending with elucidation of diverse strategies at survival, passive/spiritual, militant and pro-active; with individual case studies of people in despair and stress cocooned in-between, followed by an attempt at historical exposition. The final chapter is seen as one that offers optimism that international responsiveness and a combining of agency of governance and community can push to the background situations of stress and despair.

In the case of Thiruvannamalai, a pilgrim centre, resources may not be seen in purely commercial terms, but as something over which both a pilgrim and a tourist would stake a claim to. The struggle for the ecological survival of Arunachal mountain and the court cases which have centred on its environmental protection raise several issues that concern us. The disjuncture between the tourist and the devotee that gets brought before the courts of appeal is just one aspect of the conflict. The right of the devotee to worship in silence, that of the tourist for material comforts and that of townspeople to prosper through tourism– issues they bring to fore are not confined to Arunachal, but come out in the chapters that follow. Situated within a context of globalization, tourism, consumerism and pursuit of luxury and pleasures, the indigenous people like scheduled tribe and scheduled caste communities on the margins experience resource conflicts at different levels of intensity and meaning. The concept of 'indigeneity' itself is under contest.

Francis Kulirani's chapter on the North-East region of India which is rich in natural resources, considers tourism as a development option for the region. Many of the States in the region, land-locked and having typical geographical features have recognized it as a viable and welcome option. Kulirani suggests coalition-building of indigenous communities of India in political, economic, cultural and socio-religious levels through community development centering on tourism development.

C.P. Vinod's chapter contrarily, cautions that when migrants claim their shares of the bounty, indigeneity might lose out in the course of development. He bases his argument on the experiences of communities living in Wayanad, which like the North-East has ecological specificities that mark it off from the rest of the landscape. His argument is that a congenial physical atmosphere alone is not sufficient to ensure minimum standard of development among the communities in the habitat. Social practices of the communities are to be understood in order to explain successes and failures in the context of development. Natural bounty alone will not lead to development.

Migrants put strain on an ecosystem as in the case of Wayanad, but it does not take the same form everywhere. In Assam, 'the tea community' that Amalmani Sarma studied is constituted of almost 20 per cent of the population of Assam, numbering 70 lakh people. It is a conglomeration of labourers drawn from various parts of India. There are labourers from Chhota Nagpur, Orissa, Madhya Pradesh, Andhra Pradesh, Bihar, and Eastern U.P. working in Assam's tea plantations. They experience deplorable living situation, giving rise to political unrest in Assam with surging demands from the 'tea community' for inclusion in the Scheduled Tribes list. Conflicts between the community and other Assamese people are not infrequent occurrences. The implications of including non-tribal people in the Scheduled Tribes list on the Scheduled Tribe people who may not be able to compete with non-tribal people for resources or facilities may be a concern here while a counter argument may also be put that together with similarly vulnerable non-tribal people, the tribal people may be better able to negotiate for their joint rights with the State or the mainstream.

Vinod's and Sarma's chapters draw two different types of migrant experiences. While in Wayanad, on the one hand, migrants are usurping the tribal territories and alienating them from their own habitat, in Assam, the migrants are also working in deplorable conditions for a living while enabling the plantation owners to amass wealth. The 'tea community' has no entitlement to the State's prosperity through plantation wealth. Naturally, this resentment builds up and gives rise to tensions.

While Sarma's chapter pertains to in-migration into plantations to cater to labour demands, Rajendra Prasad discusses dislocation of the Kadar, a Particularly Vulnerable Tribal Group in Kerala, for the Athirappally hydro-electric project and eco-tourism and presents the argument that indigenous people should be given de facto decision-making powers regarding the use of the natural resources in their own habitat so that their marginalization may be averted.

Marginalization by the larger society is a contemporary reality for many indigenous populations whose habitats are opened up or invaded by more dominating populations. The stress arising from such resource conflict forms the subject matter of Bindu's chapter. Her focus is specifically on the Kurumba tribe of Attappady. An analysis of Kurumba self-perception of their stress reveals that their stress in relation to eco-degradation and loss of traditional livelihood is at the top of all their stressors, other severe stressful factors being intrusion of social evils like alcoholism, loss of traditional social control mechanisms and loss of control over the modern political system and circumstances that lead to the loss of confidence in government machinery and the State. Bindu observes that unlike many other tribes, Kurumbar do not have stress from land alienation as they are inside the reserve forests.

Two chapters on the displacement experience of the tribal people on account of an irrigation project in Wayanad district, by Mini P.V. and Prasad A.M., complement each other, but differ in perspectives in that the one by Prasad is from an insider's point of view. Mini narrates the story of the agricultural community, Mullukuruman (Mullukurumba) that was displaced by the Karapuzha project. They had to abandon their wetland agriculture and belief systems and make livelihood adaptations. In the changing scenario, their traditional knowledge has become obsolete. Prasad, an insider himself and an anthropologist, notes the injustice not only in terms of land alienation and displacement, but also in the differential compensation offered to tribal and non-tribal people for their losses due to their dislocation. In a critique of development, Prasad notes that many Mullukuruman do not stay in the resettled colonies. Besides the poor condition of the houses and the lack of opportunity to pursue the only livelihood option–settled agriculture–that their skills permit, the fragmentation of community and loss of their valued traditions trouble his people. He is justly acerbic in his remark that the development of a tribal community comes not from giving them some land but from treating them with equal justice as citizen so that they may avail opportunities like any other citizen and move ahead. He concedes that development certainly needs some sacrifice and sacrifice for common good is important, but the disturbing question is: why is it that some communities are always the chosen ones to make the sacrifice? Both State and civil society and the global community will have to find an answer to the unjust development.

Shashikumar, S.K. highlights the plight of another scheduled tribe community, the Koraga of Kasaragod that claims to be successors of King Hubashika. In their habitat, a shift to cash crop cultivation of rubber brought under rubber cultivation the hitherto uncultivated areas where raw material for basketry, the traditional occupation of the Koraga, once grew in plenty. Livelihood resources thus taken away, the Koraga's plight is precarious. Discussing the issue, Shashikumar raises an interesting argument: those communities that have acquired some social-cultural capital are able to multiply them with universalist rights, but those who have only particularist rights are placed in a situation where they have to acquire some amount of capital first in order to develop.

Subash V.S.'s chapter on the inland fishing *Padanna* community of Kerala is testimony that not only the Scheduled Tribe communities but also some Scheduled Caste communities are negatively affected by human interventions into ecology.

Construction of dams and bunds and excessive sand mining, have reduced yields of paddy and fish varieties, negatively affecting the livelihood of the *Padanna* community.

Suresh, K.P. documents the Mavilan community's subsistence skills and their superior ecological awareness and their livelihood techniques. Arrival of settlers into their habitat during the Malabar migration has caused eco-cultural changes. Pragathi's chapter highlights the health repercussions of ecological disturbances, while Mohan Kumar Bera discusses the struggles and adaptations in the Sunderbans region. The beautiful, but flood prone Sunderbans, is a challenge to the migrant illiterate people who cannot migrate again in the face of disasters. They try to adapt to the changing environment using folk tales and stories to communicate effectively among people of different levels and through a strategy of combining experience and alternative ways of natural resource utilization.

Amruth's chapter is situated at this juncture as a historical conceptualization of what he describes as the 'tribal problematic'. It pertains to a different time frame of the Travancore State and how indigenous people were conceived in official narratives and legal discourses in the context of the State controlling shifting cultivation. He contends that transformation from 'native natures' to 'modern natures' was attempted through successful deployment of modern knowledge or science such as forestry, agronomy, fishery and arts and crafts of converting nature into fine artefacts; technology and industrialization also get used in this enterprise. The agrarian traditions of most of the indigenous communities in Travancore were disrupted and they were incorporated as underpaid compulsory labour in the project of colonial forestry, Amruth concludes. Faced with disruptions to established relations through State interventions, other population's interventions or vagaries of nature herself, people have to develop strategies for survival. It can be negotiating with the State or other populations, developing alternative resources, moving away from hostile environments or finding coping mechanisms. The three chapters to follow speak of three diverse responses to the issues: Geetha Devi's chapter is on political nature of protests of local people and argues that local movements are perhaps more significant than global ones, while her notion of an eco-habitat has global dimensions as well. Sahoo and Nanda's paper and Mathur's paper are on conservation. While Sahoo and Nanda draw a picture of conservation through a local perception of the sacredness of the grove, Mathur's paper is of a well-planned environmental conservation programme with international assistance, State initiative and people's participation.

The volume brings together diverse experiences of ecology-indigeneity interfaces and resource conflicts and constraints and adaptations that call for multiple levels of understanding and strategies for conserving and promoting a healthy environment. Survival of populations, indigenous or international, depends on the success of such negotiations or strategies. Nevertheless, there is the deeper understanding that peace and sustainability of livelihood, culture and development are all inter-linked that no one can deny. As Nobel laureate Wangari Maathai noted in her lecture on receiving the Nobel Peace Prize, "In the course of history there comes a time when humanity is called to shift to a new level of consciousness, to reach a higher moral ground. A time when we have to shed our fear and give hope to each other..."

2
Foresters and New Orientations to Survival

Susan Visvanathan

Introduction

Kathleen Morrison in her essay on *Pepper in the Hills,* which is abstracted in the collection of essays on Environmental History, (*Ecological Nationalisms* edited by K. Sivaramakrishnan and Gunnel Cederlof, 2005) considers the role of tribals in the Western Ghaats in the productions of pepper so valuable for purposes of trade with the West right upto the early twentieth century. She describes these uplands people as being hunters, farmers, pastoralists, even bandits, repudiating the idea of 'simple aborigines'. She writes,

> "I will be concerned here to trace some of the changes and possible changes in the organization of foraging/ trading groups in southwestern India coincident with the expansion of the coastal spice trade and the increasing integration of this region into a world economy in the immediate pre-colonial and early colonial periods; that is between about AD 1400 and 1700" (Morrison in Cederlof and Sivaramakrishnan, 2005: 45).

Consider the contemporaneity of this interest on the website "Kerala Aboriginal Select Black Pepper." The company is from Canada and, is called "PaulsFinest.com" It says:

> Ideal for manual and electric pepper mills
> For 2 years I've been searching for a reliable wholesale source of super-premium black pepper and I FINALLY found it! As good and reliable as the Malabar pepper is that we get from the Peugeot supplier, this is even better! It's very hot, spicy, fruity, complex and bold. This is a pepper to use when you want to taste PEPPER !

Recently in Palakkad, wishing to buy pepper for friends in Delhi, I went from shop to shop, and received a low grade fungal white pepper, which I shouted about ! Later, I realized that so far all the pepper I did bring home had been from Syrian Christian gardens. Tamarind for fish, and pepper for everyday use is produced in the small kitchen gardens of relatives. How then shall we think about foresters and their present location in globalised contexts?

Tourism is one aspect; the other is survival in their habitat in the manner in which they are accustomed without being exposed to predators, who are culturally, politically or ideologically vested with powers to displace forest people. I wish to use the term forester, since this takes into account all those who make biodiversity their concern. It might be relevant here to look at the website of the Australian aboriginals and Kerala aboriginals where the similarity between them as a racial type has been communicated and endorsed. Colonial photography as well as tourist photography has raised many issues regarding the presentation of the self vis-à-vis the photographer and his or her subjects. This was a problem recently raised in Bhimtal by Kiranmeyi, a faculty member of the IGNOU in Delhi, in the context of her own work with black *ghetto* women in America with regard to hairstyling practices.

Ramachandra Guha refers to the contemporization of tribal studies by writing about the painter J. Swaminathan who was the curator of the Bhopal museum. He was a man who had admired Verrier Elwin, so he sent out thirty students to recover and learn from the natural archives of the tribals, namely their homes. "The students fanned out into the forests and uplands while Swaminathan himself headed for Patangarh. Here he found one house decorated with the most vivid portraits of birds in flight and tribal deities in a zestful mood" (Guha, 1999:323). Bharat

Bhavan built its collection around this artist, Jangarh, who was a Pradhan from Elwin's village and related to his wife Lila. In contrast, perhaps, is the story of the tribal artist trapped in a modern Japanese art gallery, far from home, commissioned to produce 'tribal art' but destined to die homesick, alone, desperate, and by his own hand.

Most of the questions of tribal culture are in the twentyfirst century, swamped around the problems of nation-state politics, arguments for integration and assimilation, quests for nation and sovereignty, and the commercialization of their territories and products through tourism, and the forces of globalization, including accessibility to territories by police, administrators, hoteliers and political parties.

When it comes to the question of the forests, tribe and caste communities may be severally divided or interlaced with shared interests. Many of the new problems regarding the reading of caste, in the case of village communities and peasantry, are to draw from political scientists and historians as much as sociologists. Sociologists have tended to stay with the categorisation of caste within field-work studies. The enhancement of the '*varna*' model in terms of the continuous reading of *jati* as the established practice for the '80s and '90s has been valuable. Village studies have been substituted by a variety of area studies where caste and class have been analysed in terms of livelihood patterns. The best example of this kind of work is MSS Pandian, whose Brahmin and Non-Brahmin has given us new indexes for reading caste politics in terms of the grammar of mobility. Similarly the corpus of work associated with Satish Saberwal has defined the varied locations of power in the last hundred years. How can we understand domination, particularly when it appears through bureaucracy and law? What happens to caste and class and religious identity when tradition is re-read through a map of colonial interventions? In this context, we may benefit from Ananthamurthy's reading of the co-existence of time, where in a public lecture given at NMML in the early 1990s, he spoke of each segmentalising his/her experience and identity. Orthodoxies and right to free choice with regard to marriage, food practices, and occupation or companionship alternatives intermesh in contradictory but not problematic ways.

Kumari Jayawardena (2007) in 'Erasure of the Euro Asian' refers to the poly-vocal voices, the blurring, highlighting, amnesias or stridency in the choices people make about recalling their past and their intentions, for the future, which are varied. Many sociologists now use narrative analyses to explain processes such as racial mixing or varieties of hybridities to understand globalization. These methodologies steeped in self proclaimed bias such as feminism or Marxism, *dalit* or right wing sociologies, all require some alertness to the matter of objectivity in social science.

Consider, for instance, Paul Gilroy's classic reading of race relations in the modern world, where questions of being what he calls 'Between Camps' is a routine aspect of existence. He returns to Fanon's preoccupation with 'epidermalised' embodiment, going on to argue that it is the reciprocity between environment and the body that we must be concerned with, where internal cannot be identified with genetic, but where reciprocity is the key ensemble of relations between nature and culture, (Gilroy, 2004:40). These are very old debates, where the ". . . .curse of racial domination is the condition, not of being black, but of being black in relation to the white ...." (*ibid.*) The *ghettos* of tribal people in reservations often resemble the concentration camps as a "novel form of political administration, population management, warfare, and coerced labour" (*ibid.*: 61).

If there is mutuality between human beings and culture, how do we understand this relationship in modernism? We know that Max Weber's methodological call to stating one's bias before proceeding to explain social phenomenon or the directive to keep politics outside the class room has been of great significance. Similarly, Geertz's call to believe in the task of an objective social science, in spite of the calling to art, creativity, spontaneity and expressivity, is well taken. Feminism has liberated many spaces of social documentation once rendered invisible or silent. Genealogical maps, questions of land ownership, *ghettos* having a majority of women as heads of households because of issues of work or displacement, all draw our attention to the problem of gender and identity.

New questions regarding urban space are generated through media analyses, which include film, photography, advertising, the World Wide Web, and new forms of bureaucracy and domination. Analyses of citizenship rights and new forms of democratic judicial enquiry are pertinent to the questions of reading urban space. One of the most interesting debates is around the problem of the peasantry as it appears in the map of urban life and factory production. Sociologists and historians have proved over and over again how the factory worker is drawn from peasant or tribal background. Who is the immigrant to the industrial belt? Special Economic Zone (SEZ) and the issue of confrontation between displaced peasantry, political parties and mafia bosses, provides many sociological issues that twentyfirst century Indian sociologists have to encounter. Exclusion refers to large masses of people being of the privileges of development, while their labour may be used to manufacture the means of production. Human Rights Watch Groups have always been concerned with the problem of the servitude of the Indian masses to the State, and yet we know that the poverty of the large mass of the Indian people is matched by their resilience and their ability to shake up government in power through the electoral processes.

Many of these questions are raised in Paul Graves-Brown's collection of essays (Graves-Brown, 2000) where the authors consider material culture as a problem not just of ideas and relationships, but with regard to objects and things, where the human body may also be seen as an artifact. Things according to him have life histories–technologies, phase of invention, commercialization and adoption. Artifacts are manufactured, used, discarded, preserved or reused. In this volume, Bruno Latour who uses the function of the key to understand what it means to be locked in, locked out or to be in a state of being controlled, or innovating ways to escape control writes that when it comes to material culture we must look at "circulations, sequences, transfers, translations, displacements, crystallization. . . ." He says,

> "Consider things and you will have humans consider humans, and you are by that very act interested in things. Bring your attention to humans, and see them become electronic circuits, automatic gears or softwares" (Latour, 2000).

This depersonalization of knowledge through the customs of war zones and post-modern worlds help us understand how individuals become mere atoms to be observed on television or computer surveillance screens, having little identity other than their visibility.

Mediation for Latour is a process of how networks are created.

In a recent seminar held at NMML, Delhi, the *Kalpavriksha*, a well-known Delhi based organization interacted with government officials, journalists, activists and university scholars. It is this intermeshing between various bodies of society that promotes a healthy and optimistic view of the world. Checks and balances, support and questions of criticism and free speech which is disciplined in its motive while being completely judgemental are the signs of a free society.

Kalpavriksh and *Grain* brought out a publication for public discussion '6 years of the Biological Diversity Act in India,' 2009 at NMML in February 2009. They write in this text that "Human interactions with biological resources are determined by cultural contexts, religious beliefs and economic considerations. The diversities of local practices through which people spiritually revere, carefully select and collectively nurture diversity, comprise the means of managing these natural life forms. These practices aim at achieving a delicate balance between the need to use nature's resources to meet absolute needs and allowing for the regeneration of natural systems. There are, however, different world views, which seek to harness natural systems often resulting in its over-exploitation. The term 'diversity' then becomes a 'resource' which can be accessed, marketed and controlled. This is where conflicts develop, between users of natural resources, precipitating at both ideological realms and practical instances."

Implicit in this Act of 2002, and the Biological Diversity Rules, 2004, are certain key dimensions which are the ban on transfer of genetic materials outside the country without permission, prohibition on claiming patents on Indian materials without permission, control of the collection and use of materials by Indians, but greater freedoms for those communities traditionally engaged with these tasks, measures for research use, joint research with consequences for monetary benefits (national interests *vs.* local interests) provisions for conservations and protection, provision for local communities to be involved in the manner of disposal of resources, protection of indigenous traditional knowledge, regulation of the use of genetically modified organisms, setting up of National State and Local Biodiversity Funds to support conservation and benefit sharing, setting up of

Biodiversity Management Committees (BMC) at local, village and urban levels, State Biodiversity Boards at the State level, National Biodiversity Board at the larger integrative level of the Indian State, etc. Kanchi Koli, presenting the report asked for the collaborative interaction of local people, NGOs, government and the intelligentsia.

I now come to regenerative forest policy and its success in the case of Tiruvannamalai.

In Tiruvannamalai, the case of peasantry and afforestation has several dimensions, for at one level, it is landless labour which is employed by the Greenings Association to plant saplings. Their wages are low, and they are local grass-cutters who are in the pay of the association. Annamalai Reforestation Society has a board consisting of several members who appear annually for meetings. Yet the volume of work done in spite of this low profile attitude is substantial. It is dependent on donations from visitors to Ramana Asramam, and the guiding light of the association is an old couple, Mahalakshmi and Suryanandan, who concern themselves with the day-to-day arrangements of the nursery and the sapling transplantation. J. Jayaraman, the librarian, has also been centrally involved, since the gift of the land for the nursery was made by a friend of the *asramam*, who inherited some money from her father, and told JJ, that she had never owned land. Once the land was bought, she transferred it to the society. In some senses, it is like a miracle.

However the real spirit behind the afforestation programme has been Apeetha Arunagiri, who has according to V.S. Mani, (one of the trustees of the Ramana Asramam) been the single person for thirty years most involved with reforestation. It is her story which is central to this tale, so I shall be using some of her letters and documentation in the web to communicate how efficacious her work has been. Presently she works with school children and housewives, mobilizing support from visitors, cafe owners, care givers of various kinds, including a man called Aran who runs an organic farm-cum-school for local children in an orphanage.

It is important to note how variegated the term 'peasant' can mean in globalised contexts. Those who stand for commerce and overbuild in Tiruvannamalai have stakes in saleable land as farmers or merchants, those who plant trees for ecological sustenance, also are tillers of the soil. Resulting from the problems of commercial overbuilding, there is tension in town, because the contestants will be set up against one another. Out of this division, arose a court case, which is also the subject of this chapter.

The Arunachala Mountainscape Tiruvannamalai, South India: An oral history inscribed in memoirs, newsletters and court records.

One of the most interesting problems in the study of the Ramana Asramam in Tiruvannamalai, Tamil Nadu, is the understanding of the substantial work that has been done by its residents. This has included everything from clearing land and constructing, and the more routine tasks of cooking, cleaning and keeping accounts. Perhaps their perspective on work may be better understood if we look at a narrative of a resident who was a companion of Ramana Maharshi. Sadhu Mouni had spent some weeks at a Catholic monastery in the heart of Paris and spoke to the brothers of this order.

> They accepted everything without question. Then I understood the meaning of 'renunciation of one's own will' and its place in spiritual progress. It was but another form of extinguishing the so-called ego. . . When I asked them which they preferred, prayers or work for the community, they told me simply that their work was only another form of prayer. (*Sadhu Mouni, 2001:57*).

Work and worship has been the key idiom of life at Ramana Asramam, the logos being 'activism and detachment' (Visvanathan, 2007). Superiority of the rational method can only be an aid to the institutionalisation of the cult, and the family of the Maharshi, who has inherited the mantle of responsibility, has sustained its sense of virtue and integrity, by providing to the community which it services, the tradition of hospitality and transparency. This works in two ways, one by providing hospitality to those who come as seekers or visitors or are in abject need; secondly by outrightly clarifying that donations are not the significant index of the *asramam's* success.

Hundreds of people mill around in the *ashram* every day. They are visitors, who live in Tiruvannamalai or outside and come everyday for a couple of hours. The most famous of these is perhaps David Godman whom everyone admires (*see* Godman, 2000).

However, since he is reclusive and reticent, and a famous writer of books on mysticism, not many people really engage him in conversation. Many of the white visitors form cliques spend time together at the tea-shop outside the *ashram* where they sit smoking cigarettes with the *sadhus* and watching the traffic, the beggars, and the grand and silent mountain. There are some who wear dread locks and communicate visually the idea that Saivism is the idea of excess. Visitors to the *ashram,* however, are encouraged to be circumspect and to be properly clad. One glance from one of the trustees is enough to communicate that someone is disorderly, and even children are disciplined regularly if they should chase the peacocks, pluck flowers and fruit or play too loudly in the hall. Yet the everyday relationship between natural things and the order of the social is what much of the published narratives around Ramanamaharshi consist of. Since much of what Ramanamaharshi said was recorded by loving scribes, we can only say that the visual evidence of the past is substantial. Ramana's family was also traditionally village or small town notaries and lawyers, so the tradition of record keeping was implicit in the family legacy.

What is interesting for the oral historian and the anthropologist is the act of inscription. Everything that is recorded is a verbatim text, which when it stands the test of time (an index of publishing history) must also be cross-checked or verified.

A lot of the conventional questions raised in the theology of inter-religious dialogue are surpassed in non-dualism with the eternal resolution of boundarylessness. This might raise existential problems for those of us who are situated in mundane worlds, but to those who enter the realm of the boundless sacred, it is a given. To believe or not to believe, is not the question, for in India, we are free to do either.

> To one seeker, Maharshi replied to the question, "I am told that according to your school I must find out the source of my thoughts", with the answer, "I have no school" (Mudaliar, 2003: 267). Looking for the source meant simply understanding who '*I am*', *t*he rest was the seed and tree story (*ibid.,* 268).

The self was real, and further "All this is Brahman", therefore, in *Advaita, Shakti* and *Shakta* were conjoined. Bhagavan

Ramanamaharshi argued that different paths developed "to meet the needs of different minds differently constituted, but they all reveal the same Truth" (*ibid.*, 270).

Sometimes, he would ask visitors the questions posed to him. When someone asked him what one ought to do for the betterment of *atma*, Maharshi posed the question back, "What do you mean by *atma* and by betterment?" To this the visitor said "We don't know all that, which is why we come here" (*ibid.*, 273).

Maharshi goes on to say, "Questions are always about things that you don't know and will be endless unless you find out who the questioner is. Though the things about which the questions are asked are unknown, there can be no doubt that a questioner exists to ask the questions, and if you ask who he is, all doubts will be set at rest." (*ibid.*, 273)

It is understood in the light of Ramanamaharshi's teaching that the contemplative path is chosen by seekers because they need to, whether they be householders or others. It is a path that is voluntary, and does not offer riches or physical apparatus of any particular convenience. What it does offer is stillness of the mind, which we call tranquillity, and the ceasing of the apparent struggle that goes into the fulfilment of duties. Maharshi's teachings were immensely practical in the carrying out of duties, so that they no longer felt burdensome, but filled the doer with a sense of that happiness, which Ramana called '*sahaja*'. There were devotees who looked for visions and palapable evidence of their years of meditation. "Visions are not a necessary stage. To some they come, and to others they don't, but whether they come or not, you always exist and you must stick to that" (*ibid.*, 275).

In this chapter, I attempt to do two things. One is to record the work of an Australian ecologist called Apeetha Arunagiri, who has laboured for thirty years to provide green cover to Arunachala, the holy hill which is the embodiment of Siva and Shakti, in divine embrace. When myths provide a mnemonic which is a natural landscape, then the orientation of the sociologist is to show how this may affect the actions of individuals. The second problem is to consider the time bound case of a court ruling on Arunachala mountain and the negotiations between various parties such as ecologists and development agents and government bureaucracy. Since court cases keep circulating in various stages of appeal and counter appeal, my data is only a representation of social relations which have a historical location. It does not attempt to keep up with the events, which constantly change.

II

In 2005, on my tenth visit to the Sri Ramana Asramam, in as many years, a woman called Apeetha Arunagiri stops me on my way. We are near the gardens where the peacocks gather, where there is elephant grass, cut everyday as fodder for *asramam* cattle. The garden overlooks a water tank, and these flamboyant but tame birds can hide from the monkeys under the lemon trees. Aruna's interests over the last thirty years have ranged from forest cover, engendering protection from forest fires, community participation, informal and formal education for children and adults alike, water harvesting and puppet-theatre. The Arunachal Kadu Siva Plantation is her creation, with the help from an Australian NGO and resource base. She writes in one of her newsletters: The AKSP report,

(*my newsletterbuilder.com*) :

> We have not had an appreciable fire for two summers, thanks to the community participation fostered very largely by our two District Forest Officers, who have recently been transferred. These gentlemen were tremendously encouraging and helpful to all of us engaged in the Greening of Arunachala; in fact we of the AKSP owe our very existence to them and have only recently realized the auspiciousness of their term.
>
> The AKSP has permission to plant on the hill within the area designated Reserve Forest, this area is off-limits to other than Forestry Department personnel unless special permission is given. Very good reasons promote this structure. Within this special permission which we have been given, we in turn honour the original, cognitively competent considerations behind it. AKSP is very happy to be planting on Reserve Forest because in this way we are enabled to work without disturbance.

In another report, (*saranam.com*) Aruna writes,:

The past two monsoons have given sinking rain and a new lease of life to the parched landscape of the denuded mountain, Arunachala, the embodiment of Lord Siva. Luxurious waterfalls have resulted in overflowing tanks for two consecutive years so we have enormous optimism for our task of greening of Arunachala.

During this 2006 season, members of the Village Forest Committee of Arunachala Kattu Siva Plantation realized that eleven streams flow down from the slopes within the areas allocated for our responsibility. After nearly twenty five years engaged in fostering the return of a forested natural lingam, I feel confident that the greening process will surmount all difficulties, it is too strong for defeats now.

A sure sign of an awakening within the community to the value of the forest is in progress presently near to our site; the renovation of three previously neglected Vediyappan shrine, carried out by the Panchayats of two adjacent villages and by ourselves.

This indicates that the wilderness is being recognized as valuable once again, as it was in antiquity. Vediyappan is the god of wilderness.

This means it is much more likely that the plantation on the mountain will be respected; now it will be protected by the psychological involvement of the community at large. Moreover our Forest Department Conservator in Vellore and the Principal Chief Conservator in Chennai are determined to put a stop to the illegal grazing wood cutting and burning which have so far thwarted our efforts until recently. We have not had a fire for two seasons now.

She writes in detail, from Sri Ramana Asramam, where the greening programme has had its greatest support,

The town of Tiruvannamalai on the east side of the mountain became District Headquarters some twenty years ago, causing an unprecedented increase in population, now around two hundred thousand. Over-exploitation of water is endemic. During these years a growing awareness of environmental mismanagement has contributed towards the present day upsurge in intention to reforest the potentially resourceful monolith.

Arunachala Kadu Siva Plantation is linked to Rainforest

Information Centre in Lismore, Australia and has taken responsibility for six hundred hectares of denuded Reserve Forest Lands on the lower slopes of the mountain. The thrust of this project is social ecology. Twelve rural people, mostly illiterate, are engaged in water conservation strategies, seed collection and nursery work, and the plantation, maintenance and protection of saplings on the slopes. Day-to-day decisions are made on the basis of dialogue reaching consensus.

The primary objective of the greening of Arunachala is the rejuvenation of our artesan system. However, the Arunachal Kadu Siva Plantation has a wider social ecological import:

- *(a)* *to undertake responsibility for solutions to civic needs;*
- *(b)* *to participate in making decisions about these responsibilities;*
- *(c)* *to utilize and regenerate indigenous knowledge;*
- *(d)* *to follow through the work beyond the call of duty; and*
- *(e)* *to communicate about its meaning on every opportunity with members of their own villages and in the wider community.*

The reforestation of Arunachala will furnish Tamil Nadu with a perfect environmental precedent site.

Because a mountain in the centre of a plain is such a conspicuous advantage in water management.

Because the terrain has been so obviously ravaged in recent centuries that a return to health will be spectacularly dramatic.

Transformation will be witnessed regularly by a vast number of persons from all over India since Arunachala has become an extremely popular pilgrimage place with an estimated six to eight million visitors every year.

The upsurge to reforest the mountain manifests in the educated culture. The Tiruvannamalai Greening Society, formed a few years ago by community members and administrative officers, has still not been formally recognized by the State Government. A recent move on the part of devotees of the mountain to seek legal protection for the sacred mountain has recently failed in the Supreme Court, paving the way for the construction of hotels and tourist facilities on private lands adjacent to the mountain.

Aruna Apeethagiri's work is formidable in terms of its consistently 'local base' stability.

III

Aruna Apeethagiri and I have spent some time talking with Captain Narayanan, a long time resident of Ramana Asramam, about the greening of the Holy Mountain. He tells us that he has spent many years fighting for the ecological survival of the hill, and has been a participant and witness to the court cases which have centred around the environmental protection of the mountain. They were successful in some things, like making sure that the construction of the administrative officer's and the Commissioner's Bungalow did not take place on the *pradakshina* (circumambulation route) but the residents of Tiruvannamalai are hostile to any suggestion that no construction work of commercial buildings should takes place on this route. Over the time, the *asramam* has lost its ability to hold out, and has taken a more submissive position, apparently. It is well known that the town has attracted millions of people over millennia. Its magnetism is both mythic and therapeutic. I believe this is one of the great sites of Indian religious and cultural heritage and should have the status of a World Heritage Site. As Captain Narayan, a retired ship's commander, hopes their plea that the Arunachala will receive this status in time be heard, so do I.

Commerce and Tourism are the catch-words of the twentyfirst century. We cannot deny the impulse to prosperity that the people of a small town feel. How to safeguard the interests of the people is the task of oral historians and sociologists and policy makers. My suggestions are that all small towns which have benefited from commerce and tourism have actually merged with the ecological patterns of the terrain. Hotels and shops should be concentrated in a new quarter away from the site of antiquity and conservation. The temple and the mountain itself can allow for the cohabitation of ASI and local merchants (as in the case of Rome and Delhi). The 14 km walk around the sacred hill may be kept free as a trust site for present day ritual purposes, as well as, for the travellers and tourists of future generations. The percolation of the benefit from tourism should be made visible to the people by way of jobs, schools, medical facilities and small town necessities for recreation and learning. Garbage collection is now better organized, but workers still do not have gloves to scoop up the muck from open drains.

Banking is perhaps easier now; I do not hear complaints from foreigners. Search for the mysteriousness of God goes side by side with the mundane aspects of survival. The class and comfort interests of foreigners seem palpably the new aspect of trade in Tiruvannamalai. From being a small town, quiet almost rural, it is now a bustling tourist town. Nameless as devotees wish to be in their service of *Bhagavan*, for the sociologist personal biographies and official histories are interesting juxtapositions to the telling of a story.

IV

At the *asramam*, it is understood that Ramana's teaching participated in in the sense of the experience of the godhead, that ignorance is removed because the mountain Arunachala liberates, that one would not be there but that one is a seeker. Because it is a congregation of seekers, the experience of bliss is commonly understood among those who gather here. Those who know of this describe it as a calmness of the heart and mind, a tranquillity that forces them to return or hope to return. K. Swaminathan, a biographer of Ramana and the editor of Gandhi's collected writings in 100 volumes, described the wanting to return to the *asramam* to see *Bhagawan* as so strong a desire that he wished he could be tied down by elephants.

When I first visited 'Thiru', as it is called by frequent visitors to the town, it was a 'small' town. Now it has burgeoned by the influx of pilgrims (5,000 visit the temple everyday, according to the local chemist who has a shop next to the Arunachala Temple) and the numbers multiply geometrically as the years pass. It seems to have become the hub of priestly and commercial activities. Television has contributed much to the missionising appeal of Saivism. Descriptions of the market, temple and pilgrims remain fairly constant in the anthropological literature whether it be

Christian, Muslim or Hindu. Religion and commerce, with the possible yeast of politics, makes the town what it is. In Tiruvannamalai, there is faith, but fortunately no communalism.

To me, as a theoretician of dialogue between religions, non-dualism is the answer to varieties of discord which we see as everyday occurrences in a globalised world. Non-dualism is the space in which everyday pluralism surfaces. It is an existential quality, which allows the myriad nature of the world to be ever present. As individuals we enter a space of *communitas* and this involves both respect and sharing.

On the other hand, technological imperatives (such as the megaphone) can be very rude interruptions in the quiet and sacred spaces that Nature allots us, and which is confirmed by cultural or theological formulae. Tiruvannamalai, eleven years after my first visit in 1996, provides such a sense of shock and rupture. Trucks bringing agricultural produce and other merchandise, buses bringing pilgrims by the hundreds veer through a traditional street with enormous noise in its wake. The gods during carnival time are taken around in tractors which spew the most horrible smoke. Pollution levels, on account of diesel and firecrackers during the festive occasions have to be seen to be believed.

And everyday is a festival day, in this ancient town, because everyday is a holy day. Everyday brings to pilgrims, according to the legend and to the faith, honours and rewards, both material and spiritual. And who can be excluded? The numbers only increase. As a result there are problems of discipline and order. Residents of the Ramana *asramam* are forced to become aggressive bouncers who keep the crowds at bay, using only their eyes, linked arms or speech to keep the crowds from hurting each other or damaging the shrines. Women who visit Ramana *asramam* to meditate say that young men in the crowds swirling outside at *deepam*, the annual core festival of Tiruvannamali are lascivious and eager, and hands reach out to touch women's bodies. The swirling river of people is charged with emotions which are very mortal.

A great deal of Durkheimian sociology is about the size, density and violence of the city, and how does this affect the social interaction. At the Ramana Asramam, residents are able to cope with the greatness of numbers by coping with their needs pragmatically. Every year, some 5,000 guests stay in the *asramam* rooms, according to Dr. Murthy who is the keeper of ledgers and keys. During festivals and Sundays, day-trippers constantly float in and out, for a brief view of Maharshi's *samadhi* and garden. The bookshop remains open all night, and the monks and householders looking after it never sleep, it seems, as the flow of the people is continuous. The cooks are kept busy without a day's break, and seemingly without any holidays, in the service of *Bhagavan*. Pilgrims have now started celebrating their personal functions and commemorations of their own birthdays and anniversaries by coming to the *asramam* and asking for special favourite foods to be cooked. It has the appurtenances now of an industrial kitchen, and to cope with the crowds, sometimes the priests help the servers. They have to run with heavy buckets of rice, *sambhar*, vegetables, *rasam*, buttermilk, *payasam* among an entire convoy of pilgrims. On festival days and weekends, 300 to 500 people are fed in twenty minutes, before the second batch is called in. The second batch consists of Ramana office staff and workers, and the atmosphere is easier and colloquial. There is a general camaraderie, food is sometimes short, and buttermilk watery, but the workers eat heartily with conversation and laughter and an unself-conscious joy in the everydayness of their routine obligations and duties. While *asramam* guests are fed on washed banana leaves, workers and residents have stainless steel plates (ever silver is the local name for these plates) and only if they are not entirely integrated as wage workers, they are given banana leaf or stitched leaf plates. Service *vs.* wage work is a very sharp divider. Those who serve in the *asramam* often come from wealthy households, bringing their capital with them as donations to the *asramam*. They are known as *sadhakas* who attempt to attain the light which is beyond all thought, all events.

When there is doubt about achieving the light the answer always, in Ramana Asramam, is to follow the eight-fold path, which includes celibacy, vegetarianism and *yoga*, and the goal will be achieved. These are people who are permitted to renounce the world and so become its ornaments. Some find it as *karma-yogis* transcending the dualism of thought and action.

For a sociologist, what is not difficult to imagine, are the routines, and the symmetry arising from these. The ethnographic present, which decades later may be read as a historical record that achieves certain symmetry when it draws from tradition. The everyday tasks are set out and are immediately comprehensible.

Times of meditation, community prayer, chanting, meals, work, social interaction and silence, all of these are precise. Yet the clock time of the factory does not approximate the rhythm of *ashram* life. As numbers increase there is a strain to conform to the ethos of service and the velocity of demands. Each visitor feels he or she is unique, looks forward to comfort, quiet and aesthetic pleasure. Learning the roots of detachment is linked to various other tasks which include conformity to the values of the *asramam*—the penultimate rules are silence, meditation and work. Ramana Maharshi never believed that idleness was the path to spiritual experience. Meditation was a part of everyday life, and while some were set apart for this, the risks were always elaborated. "Discount all visions as distractions" was Maharshi's primary advice to novitiates to continue meditation. There is no reason for those of us who have particularly stressful lives in war torn cities to believe that meditation will not lead to calm lives, and the ability to carry out our duties whatever these are.

V

Maharshi always said of Nature, that manifested or unmanifested, we as human beings shared in the essence of the largeness of its existence. Much of how we read Nature is in the animistic terms of poetry and love. Durkheim read this in the metaphors of power magnetism, current, force, all of which he saw as common to Religion, Magic and Science. In the post-modern world, the boundaries between these are again blurred.

To understand Nature in terms of the mystery is part of the quest for solitude, whether it is in Religion or Science. Love is part of this quest. Thomas Merton writes, "Identification by love leads to knowledge, recognition, intimate and obscure, but vested with an inexpressible certainty known only in contemplation" (Merton, 1975:68).

This search for solitude is represented in the unity of the self.

> To love solitude and to seek it does not mean constantly travelling from one geographical possibility to another. A man becomes a solitary at the moment when, no matter what may be his external surrounding, he is suddenly aware of his own inalienable solitude, and sees that he will never

> be anything but solitary. From that moment, solitude is not potential–it is actual (*ibid.*, 79).

He goes on to say that a "merely subjective and inward solitude, the fruit of an effort at interiorisation" would not be enough. It has to be a "communion in something greater than the world" as great as Being itself, in order that in its deep peace we may find God" (*ibid.*, 83).

This is we know, very often the reason and the site of abstract and mystical poetry.

But Merton recognizes that the true solitary is one who seeks, and would presume any attempt to distract him or her from this path as futile.

And yet, when we create the material conditions of that solitude, we stand the chance of losing it.

One of the problems of populism is the ability of people to believe that the space they occupy belongs to them. Much of the questions raised by Social Anthropology is about how we look at Trusteeship. The definitive nature of the organization of space is a legal question, which is in the purview of the Forest Department, the Archaeological Survey, and Municipal offices and so on. Having delineated the path of silence as integral to the location of Arunachala as a historical site, which communicates that mysticism is a social fact, how can this silence be protected?

How increased numbers of pilgrims affect the town has already been mentioned. In the *ashram*, monastic life is continually under threat. Dr Murthy, keeper of the keys, tells me that he keeps out those people with money who want to use the *ashram* as a holiday resort, people who are eager to meet friends and relatives for festive occasions like New Year. This is complicated by the fact that each language group has a new year quite different from others complicating festive and ritual life. . . thirty per cent of *asramam* visitors are for instance Telugus who have now built their own imposing guest house, just across the road, facing Arunachala, since *asramam* cannot accommodate their number and the frequency of their visits and their desire for a longer stay.

There is an excerpt in Merton's work from the writing of Barry McLaughlin in nature, grace and religious development,

> Yet one of the surest signs of the resolution of the identity

> crisis is an increased capacity for being alive, for being responsible for oneself. The gradual process that will end in perfect identity involves an awareness of the fact that there are decisions in life and aspects of life's struggle that a person must face alone. . . . And as a person in this formative isolation becomes more able to appreciate the moods and feelings of others, he also becomes more able to have meaningful relationships with them (Merton, 1980:42).

'Togetherness' and 'Herdism' according to Merton is not what a community is all about. To love, one must respect the other; one must comprehend the need for solitude. To incorporate, to merge, to present the ideas of 'family' 'brethren' or 'community' as ways to integrate the solitary who has his/her own path is a common method.

Yet Merton has a fear of those who take to monasteries, the 'emptiness and angst' of depersonalization.

Merton contrasts what he calls normal life with the life of solitude. 'Normal' life is about comfort, reassurance and diversion.

> If we fulfil the role imposed on us by others, we will be rewarded by approval. These roles impose definite limitations, but in return for accepting the limitations we enjoy the consolations of companionship, of understanding support and so on. We are made to feel that "we belong" and are, therefore, "all right." What the seeker of solitude does is to renounce these, the monk is someone who wanted to "look into something else (*ibid.*, 100).

The comfort and joy of married life, conversations and friendship with others, a place in the city and commerce, all these are given up by the monastic. Yet the interesting thing about Raman Asramam is the manner by which non-dualism in a rule-bound world, allows people to follow their heart. Householders are valued members as are celibates. The householders, however, are there to follow their own *sadhana* and to support the *sadhakas* in their search. Thus many of the trustees of Raman Asramam have been married persons, such as the brother, the nephew and grandnephews of Ramana Maharshi, who with their wives have kept the lamps of the monastery burning, and who understand that :

> *The monk is a man who in one way or other pushes to the very frontier of human experience and strives to go beyond, to find out what transcends the ordinary life of existence* (*ibid*., 101).

Aware that man is somehow sustained by a deep mystery of silence, of incomprehensibility of God's will and God's love—the monk feels that he is personally called to live in more intimate communication with that mystery. And he also feels that if he does not respond to this summons, he cannot be happy, because he cannot be fully happy with himself (*ibid*., 101). It is this which Ramana Maharshi called '*sahaja*' whether for the monastic or the householder, and where I would conclude with Merton that detachment is the lot of the social scientist as much as the mystic, and that we must be careful that our techniques do not lock us in (Merton, 1980:103).

The world is a complex web of loves, hates, joys, greed and hope and we traverse it in our different ways, always hoping for that reconciliation which will allow the earth to prosper. It is in this context that I raise the problem of the court case which pertains to the greening of Arunachala. I place it here, for purpose of record, for the battle to protect the holy hill has been in the forefront of many devotees' lives.

VI

The Case is known in the Supreme Court of India, as the "*Civil Appellate Jurisdiction, Nos. 12443-47*" and it arose out of the final judgement and order dated 11.5.2001, passed by the High Court of Judicature at Madras in writ petition No. 17109/97, 7396, 7397, 7398 and 7400 of 1998.

The petitioner was the Commissioner, Tiruvannamalai Municipality and the respondents were Arunachala Giri Pradakshina Samithi and others.

The case in the Madras High Court centres around the fact that the Government of Tamil Nadu had provided guidelines for the preservation and maintenance and development of 38 temple towns, by G.o.Ms. No. 163, Rural Development and Local Administration dated 6.7.93.

In 1997, writ petitions under Article 226 of the Constitution of India were filed before the High Court of Judicature at Madras, by the respondents as a Public Interest Litigation. "It was alleged that the public in general and the followers of Lord Arunachala and innumerable persons who go around the Hill found that there are encroachments, unauthorized constructions, deviations, illegal felling of trees, destruction and devastation of water tanks. Caves, temples, *ashrams* and other religious places in the Hill Arunachala, and in the Holy town of Thiruvannamalai, and the monuments have to be protected, by removing the encroachment, by restricting and regulating the constructions. It is submitted that the entire Holy Pathway and Holy Hill be protected, and requests various reliefs/directions against the petitioner and various public authorities".

In March 1999, the Municipal Commissioner argued that there was a bias in the manner of description. "It was submitted that the first respondent has invented the Holy Pathway and the Holy Hill Arunachala afresh without any basis or record. It was submitted that the petitioner as a Civic Body is taking action to control development." The petitioned argued that there were no temple encroachments at Arunachaleswara temple except the shops which were permitted by the temple authorities. The other petitioners against the Arunachala Giri Pradakshina society and others were the Aathaiandal Village Panchayat, Adi Anamalai Panchayat, and Adaiyur Village Panchayat; the Hindu Religious and Charitable Endowment also filed its oppositions against the respondent.

In May 1999, the High Court appointed an expert body to provide a report to the court with suggestions regarding maintenance and preservation of the heritage town of Tiruvannamalai and the pathway around the hill. Mr T.S. Arunachalam, Acting Chief Justice of the High Court Judicature at Madras was appointed as Chairman of the Body. Justice Arunachalam was to be assisted by the District Collector, who was to be the Secretary of the Expert Body, and also by an architect, an IAS officer, an archaeologist, by Conservator of forests, by Secretary to Government, Municipal Administration and Water Supply Department, by the PA to the District Collector, by the Director of Town and Country Planning, by the Director of Municipal

Administration, by the Executive Officer and Joint Commissioner of Annamalaiar Temple, The Municipal Commissioner, The Executive officers, Presidents of Village Panchayats of Anapuandan, Athiuandal, Adi Annamalai, Vengikul and Adaiyur. Also by the Executive Engineer, Tiruvannamalai, Annamalai Reforestation Society, Executive Engineer, State Agricultural Engineering Department, District Revenue Officer, Revenue Divisional Officer, Deputy Commissioner, HR and CE, Engineers with Highways and Rural Works having jurisdiction along the Sreepradakshina Salai and the surrounding Regional Transport Officer.

There were in the Expert Committee representatives of important *asramams* such as one representative each of Yogi Ramsurat Kumar, Ramana Maharshi, Seshadri Swamigal and Gowthana Ashrams. Three prominent public persons who had an interest in conservation and maintenance were to be nominated by the expert body.

The records of the case describe the consequence of setting up the expert body thus:

> 8.12.'99. The expert body exceeded its scope and gave its report to the High Court and suggested that a single authority should be created, which should have power to implement the plan not only in Thiruvannamalai town, but in all other Panchayats falling within the limits of the Master Plan on the lines of the Chennai Metropolitan Development Authority. A District Heritage Conservation Act (Bill) to ensure protection and restoration of all heritage buildings, precincts and areas, until the Bill becomes an Act. The Town Planning Authority will act on the basis of recommendations of the District Heritage Conservation Committee, in all matters relating to heritage. A Heritage Conservation Fund for financing projects identified by the District Heritage Conservating Committee has to be set up and the Committee can decide what will be the sources of this inflow for this fund. The Expert Body suggested in all 29 suggestions. In all it suggested changes in existing building rules, provision of some basic infrastructure.
>
> 29.2.2000. The Report of the Expert Body was placed

> before the petitioner Council pursuant to the order passed by the Hon'ble High Court.
>
> 2.5. 2001. The Government of Tamil Nadu filed its objection to the draft scheme stating that the present scheme is against the Constitution of India. It tries to set up an executive authority endowed with enormous powers and which encroaches into the powers and responsibilities of the Constitutional Body. On the whole, the scheme appears to allocate all land use, water and sanitation functions to the committee. As per the XII Schedule of the Constitution of India, these are functions of Municipalities and they cannot be transferred to a Committee, which has nominated members only. The scheme is both impractical and unnecessary to expect all Departments of Government to function under the directions of this Committee. The Committee cannot be allowed to function as an organization which is beyond existing legislation and deemed to function like a super power.

Clearly, there is a substantial shift in this note, from a position reached in the Madras High Court, where respondent and petitioners accepted that they had interests in common, which was the protection and maintenance of a holy town. Here questions of authority and hierarchy are being raised, and while the Expert Body consisted of members who were specialists in their respective fields, belonging to the government, both State and Central, along with towns people, and *ashram* delegates, *the actual praxis of power* is the problem now. *Can the offices of government be supervised and directed by a committee proposed by the judiciary*? The opposition to the committee is put forward very sharply in the court record :

> 30.3.2001. The petitioner Municipality filed an affidavit stating that the proposed scheme authorizing so much powers to such committees will make serious inroads into the powers of the Municipal Council and the local authorities. The Committees proposed are contrary to the provisions of the Tamil Nadu District Municipalities Act 1920 and these Committees will run parallel to the Municipal Council and the same is against the provision of Article 243 of the Constitution of India.

The assumption that the Municipality makes as petitioner is that the High Court has exceeded its jurisdiction and taken on a privilege that is associated with Executive, Legislature and Judiciary.

Interestingly, the court proceedings in the High Court of Judicature at Madras, on Friday 11th May, 2001, has a preface: Order passed by Justice Jagadeesan, Hon. Justice E. Padmanabhan states:

> The Law in the past was directed unconsciously and it should now be directed consciously, to the same end as custom, manners, religion and other types of social control, namely with the preservation, furtherance and transmission of civilization.

It is this conflict between governance by judiciary and governance by State institution that sets up the dialectic. Public Interest Litigation, according to one of the lawyers who fought the case for Arunachala Pradakshina Samiti, is always vulnerable at a time when individuals may have honourable intentions but not the staying power to keep with a case till the end.

Since it was a matter of encroachment and over-build, all those who had a commercial interest in construction became offensive, and Ramana Asramam was visited by a large number of individuals, saying they were towns people, and if Asramam did not withdraw its case, there would be trouble. According to Arunagiri, the Australian ecologist who has lived in Tiruvannamalai for thirty years, it was very frightening for residents. I quote from her letters to me:

> A recent move on the part of devotees to the mountain to seek legal protection for the sacred mountain has recently failed entirely in the Supreme Court, paving the way for the building of hotels and tourist facilities on private lands adjacent to the mountains. Someone said the opposition of the town was so tremendous, that the ASI did a deal with them: they would stand back if the Municipal Council started to keep the town clean.

The real question was of livelihood, and if the pristine quality of conservation was to be put into institutional space and its guardianship assured, then a great deal of how people actually made a living from the merchandise of objects, for sale to pilgrims would decline. Curiously the petitioner in the Supreme Court, the Municipal Commissioner, argues that pilgrimage cannot be confused with tourism, and that Tiruvannamalai is a sacred town, where the livelihood of people is at stake. He describes archaeological survey operatives, or tourism development 'intrusion' as having to do with visitors who are not pilgrims. That is not his concern. He also provides the Supreme Court with photographs of Tiruvannamalai at the time of day, in the morning, when there are few people on the street, and the roads cleared off garbage, to show how well the Municipality deals with crowds and refuse!

In the Madras High Court, the Municipality had stated that :

> It was bound to grant permission for the construction under the provisions of the Tamil Nadu District Municipalities Act 1920 and the rules framed thereunder. There is no authority whatsoever to prevent the construction on private lands adjoining the *giripradakshina* pathway and beyond the 50 metres of *giripradakshina* pathway and also there is no authority to limit the height of construction not exceeding 6 metres either adjoining the *giripradakshina* pathway or in the remaining portion of the path and up to 50 metres beyond the giripradakshina pathway against the principles contained in the Heritage Act and the guidelines issued thereunder in G.o.M.'s No. 22. MAWS Department, dated 30.1.1997.

The Municipality argued that the shops are permitted by temple authorities, and that the Municipality is discharging its duties.

The Honourable Judges wrote:

> Admitted facts and stand of petitioners and respondents. On the facts of the case it is clear that there is no controversy, besides it is almost admitted by everyone, including the respondents about the existence of the Holy Hill Arunachala which has been personified as Lord Shiva and the existence

> of *giripradakshina* path over which millions go around the Holy Hill, commencing from sixteen pillar mandapam of the eastern entrance of the temple, and ending with the very same sixteen pillar mandapam. Everyone concerned whole heartedly supported and expressed that either the State or Central Authorities should step in to regulate the entire area, remove the encroachments, regulate the construction of buildings so that visibility of the Holy Hill is not affected when it is viewed from any corner of the *giripradakshina* path, besides the requirement of the necessary basic amenities or facilities to pilgrims and the villages and the residents who reside on either side of the *giripradakshina* path as well as the Tiruvannamalai Municipal limits. Everyone anxiously expressed that the entire pathway should be maintained free from all encroachments whatsoever, and all the obstructions and encroachments between the pathway and the Holy Hill have to be removed and regulations have to be brought in with respect to the height of the building or location of the building and as to the using of the space and leaving of vacant place abutting the *pradakshina* path on both sides and with respect to various measures to be taken in this respect to preserve the Holiness and sanctity of the Holy Hill Arunachala.

The Honourable Judges asked the State Government to follow up the consensus reached by both parties that the Holy Hill needed protection.

> During the entire hearing everyone has supported the move, but only had different views as to the steps to be taken or extent to which restrictions are to be imposed and with respect to the details of restriction to be imposed and the regulations to be followed. In fact we called upon the learned Additional Advocate General to get instructions as to whether the State Government is in a position to take up immediate legislation for the area in question which would give quietus to all the issues. But the learned Additional Advocate General on instructions expressed the State government's inability to bring forth a comprehensive legislation and the response of the State

> government so to say is not only lukewarm but it has failed to comprehend the problem and it had failed to respond.

(from file titled, Supreme Court of India, Civil Appellate Jurisdiction, Nos. 12443-47 Commissioner Tiruvannamalai *vs.* Arunachal Giri Pradakshina Samithi and Others.)

The judgement continues:

> However, in principle though the State government agrees they are not willing to take up the appropriate legislation, and time granted in this respect also had not elicited any response. In the light of the said background this Court on 4.5.1999 passed a detailed order with the consent of the petitioners as well as all the respondents and everyone concerned. The said direction is a consent direct as everyone agreed that *giripradakshina* running to the length of 16 km or thereabout from Anmanalaiyur Temple proceeding along with Tiruvoodal Street, Sengam Road, Pei Gopura Street, and Soma Vara Kula Street etc., and ending at the same starting point of Annamalayar Temple has to be cleared from all encroachments and unauthorized constructions and a comprehensive scheme has to be framed for the area and *giripradakshina* Salai in the larger interest of the public as well as in the interest of residents of Tiruvannamalai Town as well as villages in and around the hill (*ibid.*, 42).

The judges reiterate that both parties are agreed on the sanctity of the hill. This very sanctity will be disrupted by unregulated haphazard construction and a health risk to all. They note that :

> Mr Sriram Panchu, Senior Counsel appearing for one of the writ petitioners persuaded this Court to frame a scheme so that the entire locality could be protected not only from encroachment, unauthorized construction, deviations, but also the cleanness of the town and environment could be saved from further deterioration (*ibid.*, 42).

In furtherance of the objectives resolved in the consensus, the Hon. Judges, requested a study of Tiruvannamalai which would be presented by an Expert Body. One of the recommendations was certainly forbidding megaphones in public places! The judges did see that the townspeople would have to make sacrifices for the greater good (*ibid*. 78).

The Arunachalam Committee gave an extremely detailed and constructive report, and as a memento to town planning, it is a valuable record for urban historians and sociologists. The Tiruvannamalai Municipality reacted sharply. The court records state:

> In fact the counsel on record for Tiruvannamalai Municipality attended all the hearings but did not object to the proposal to frame a scheme. The counsel for Municipality, the State, the Centre were called upon to get instruction from the respective authorities, whether they are willing to take comprehensive measures, but all of them despite opportunity failed to avail the opportunity, which would exhibit their failure to discharge the public and statutory functions. Further, all of them also represented that this court may frame appropriate scheme. In the light of the above, we are unable to appreciate or sustain the belated objection raised on behalf of their municipality as if their power or jurisdiction is being reduced. Factually, it is not so as none of the powers of the Municipality as conferred by the District Municipalities Act or the Constitutional provision are being interfered since the provisions of the Scheme is in addition to the statutory provisions. We overrule the objections and hereby frame the Scheme appended herewith which shall be implemented by everyone (*ibid*., 79).

When the case went to the Supreme Court, the Municipal Commissioner basically communicated that it was not his place to take orders from a parallel body, the Expert Committee, which presented itself as a super power. The Municipality's jurisdiction had been replaced. When the suggestion that the Central government believed that heritage and tourism were important issues, the Commissioner made a distinction between tourists and devotees, saying that their needs were different. The violence of the agitators (who wished to build guest houses and adjuncts for commercial reasons,) when they appeared as a violent crowd, according to informants, at Ramana Asramam, was such that the PIL was withdrawn. In October 2007, an informant at the *asramam* said jubilantly, that the ecological struggle was still on!

One resident of Ramana Ashram told Aruna Apeethagiri and me that:

> *"At least the Municipality did not get to build their offices on the Chengam Road, as they wished,(the circular Road outside) otherwise imagine the bottleneck that would have been created, with people from all over the hinterland appearing with their papers for audience with the Commissioner, and jostling with pilgrim on this narrow road!"*

What is interesting for me as a reader of these court records is how much of faith is captured even within the secular jurisdiction of the courts. The litigants, lawyers and judges accept the sacred manifestation of the divine, and the right of the people to worship and live in a territory which is infused by the heady sense of myth, which communicates both love and *maya* as central theological concepts.

REFERENCES

Arunagiri, Apeetha (a), 'The *AKSP Report*', available at *http://www.mynewsletterbuilder.com/tools/published.php?action=view&newsletter_id=1409585990, site accessed on* 10th July, 2009.

Arunagiri, Apeetha (b), 'The Arunachala Kattu Siva Plantation', available at *http://www.saranam.com/aksp/index.htm*, site accessed on 10th July, 2009.

_______ (c), 'The Arunachala Kadu Siva Plantation', available at *www.ser.org/iprn/pdf/Apeetha_Arunagiri. pdf*, site accessed on 10th July, 2009.

Case, Supreme Court of India: 2001 Civil Appellate Jurisdiction, Nos. 12443-47.

Cederlof, Gunnel and K. Sivaramakrishnan (2005). *Ecological Nationalism*, Permanent Black, Ranikhet and Delhi.

Gilroy, Paul (2004). *Between Camps: Nations, Cultures and the Allure of Race*,

Routledge Taylor and Frances, London and New York.

Godman, David (2000). *The Power of the Presence*, Avadhuta Foundation, Pondicherry.

Graves-Brown, P.M. (2000). *Matter, Materiality and Modern Culture*, Routledge, London and New York.

Guha, Ramachandra (1999). "Savaging the Civilized" in Verrier Elwin (ed.), *His Tribals and India*, Oxford University Press, Delhi.

Jayawardene, Kumari (2007). *The Erasure of the Euro-Asian, Recovering Early Radicalism and Feminism in South Asia*, Social Scientists' Association, Colombo.

Koli, Kanchi and Kalpavriksh (2009). *Report of the National Dialogue on Six Years of the Implementation of the Biological Diversity Act* at a Conference organized by Kalpavriksha, Grain and the National Forum for Policy Dialogue, at the Nehru Memorial Museum Library (NMML), Delhi, February 3.

Latour, Bruno (2000). "The Berlin Key or How to Do Things" in P.M. Graves-Brown (ed.) *Matter, Materiality and Modern Culture*, Routledge, London and New York.

Merton, Thomas (1975). *The Asian Journal of Thomas Merton*, New Directions Publishing, New York.

_______ (1980). *Seeds of Destruction*, Ferrar, Straus and Giroux, New York.

Morrison, Kathleen (2005). "Environmental Histories, the Spice Trade, and the State in South India", in Gunnel Cederlof and K. Sivaramakrishnan (eds.) *Ecological Nationalisms*, Permanent Black, Ranikhet and Delhi.

Sadhu, Om (1997). *The Path of Sri Ramana*, Pondicherry: Sri Ramana Kshetra, Kanvasha Trust.

Visvanathan, Susan (2007). *Friendship, Interiority and Mysticism*, New Delhi: Orient Longman.

3

India's 'Look East' Policy

Unlocking the Landlocked North-Eastern India to the South East Asian Tourist/Trade Hub: Dimensions in Anthropology of Ethnic Tourism

B. Francis Kulirani

Introduction

The concentration of communities having Scheduled Tribe (ST) status in the North East Region of India is a known fact. It is also a fact that North East has taken the character of a socio-political entity (Shimray, 2004). The data available with the Anthropological Survey of India estimates that there are about 305 communities in the North East and 166 of them are Scheduled Tribes. In some of the States, like Meghalaya, Nagaland and Arunachal Pradesh, majority of the communities are Scheduled Tribes. The developmental achievements of the region, in other words, are directly proportionate to the development achieved by the tribes of the region.

The region is richly endowed with natural resources of high rainfall and water, fossil fuel, coal, uranium, lime stone and unique flora and fauna, but the region is struggling to evolve the right developmental strategies. The problems of underdevelopment experienced by the communities of the region are similar, in terms of poor communication, poor health services, unemployment, inter-ethnic hostilities/clashes and militancy. The vast amount of plan resources is pumped into the region for the development of 'the land', and not of the communities who perceive that the resources of the region are siphoned away by the outsiders, be it coal, oil, hydro-electric power, or lately uranium. The North East region of India comprising of eight States (Assam, Arunachal Pradesh, Manipur, Meghalaya, Mizoram Nagaland, Sikkim and Tripura) forms the homeland of 166 ST communities who form 57.3 per cent of the region's population (Census, 2001).

They are of Mongoloid stock and speakers of Tibeto-Burman, Tibeto-Chinese, Kuki-Chin and Monkhmer languages, who had specific territorial areas of control till the British Colonial annexation of the region in the mid nineteenth Century (Bhattacharjee, 1977). The Colonial administrators demarcated tribal inhabited areas as 'excluded areas'–excluded for the outsiders or non-tribes. In the independent India, ST communities of some of the States of the North East Region were also granted special constitutional protection under Article 244(2) known popularly as the Sixth Schedule (Gassah, 1997). The provisions of this enactment provide exclusive rights to the ST groups over their land in the defined areas and outsiders including tribes from other regions cannot buy and own land in the notified area; further the tribal customary laws and their political institutions are legally recognized.

Landlocked North-Eastern India

One of the reasons that can be sited as the cause for the state of affairs can be the 'Landlocked nature of North-East'. Prabhakara (2004) has examined the multi-dimensional aspects of landlockedness of the North East region. According to Prabhakara, 'unlike the other landlocked States of the Indian Union, the States of North East are not surrounded by Indian territory, on the contrary, the North East is almost entirely surrounded by foreign countries, which is a unique locational feature. Further, an even more unique aspect is the fact that 'each of these States are also internally locked, themselves locked and locking out others, unable to connect with each other physically in terms of poor transport links and more seriously, unable to make connections intellectually and emotionally with their closest neighbours or even with and among their own people' (*ibid.*). The reason for this can be found in the region's political history.

The present landlockedness of the region started in the

nineteenth century, with the British Colonial rule enforcing barriers on the old trade routes. The Indian Independence in 1947 and the formation of East Pakistan (now Bangladesh) further reduced connectivity except for a narrow corridor from West Bengal State in the north west of the region. The region having an area of 2,55,000 sq. km. has international borders with Bhutan

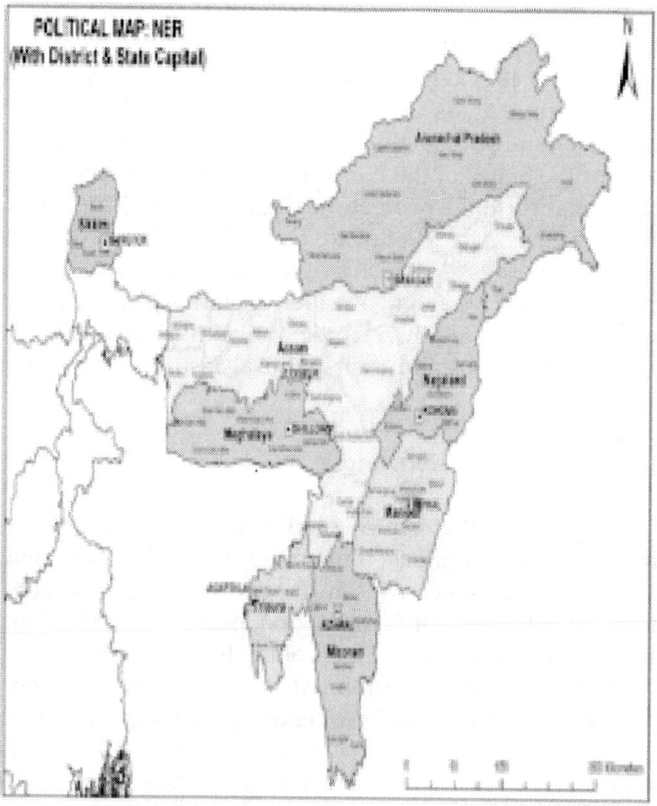

Fig. 3.1: Political MAP: NER (With District and State Capital)

and China in the north, Myanmar in the east, and Bangladesh in the south.

North-Eastern India and Challenges of Development

The concept of development in the context of distinct community identities of North East, who are mostly Scheduled Tribes, need to be examined holistically. The very formation of the States of the North East region itself was based on the dominant features of ethnicity and language. The ethnic differences and political aspirations within the States are so divergent that there is a perpetual state of ethnic unrest within. Intra-regional disparities of growth and development between ethnic groups in the plains and hills are also evident. Sharing of land resources, sharing of funds for regional development, sharing of employment opportunities and sharing of power and authority, all lead to claims and counter claims. At the same time, internal revenue generation within the States and per capita income at the level of individuals also remain low.

NER Development

The region is lagging behind in many of the developmental parameters and, hence, the ethnic unrest is currently being attributed to lack of avenues for participation of the people of the region in the development process. Several models of planning have been tried out in the region from the 1940s. In terms of paradigmatic shifts of these policies, Jayaram Ramesh (2005) chronologically categorised them as the 'culture paradigm' till the mid 60s, followed by 'security paradigm' that lasted till the early 70s which was replaced by the 'politics paradigm' till the early 80s and lastly the 'development paradigm' which is continuing to be the main policy thrust for the region from the 1980s.

Look East Policy (LEP)

The current thinking of the Government of India spearheaded by the Ministry for Department of North-Eastern Region (DONER) is to position 'North East India in a new Asia' and has come to be popularly known as the 'Look East Policy' (LEP) coined by the former Prime Minister Narasimha Rao. The LEP envisages that the NER not as a periphery of India, but as the centre of a thriving and integrated economic space, linking two dynamic regions with a network of highways, railways, pipelines and transmission lines criss-crossing the region. In the pursuit of the above policy, the DONER Ministry had organised a four-day 'North East India Trade and Investment Opportunities Week' at Bangkok in the first week of October 2007. Considering the region's huge eco-cultural capital, tourism in NER is being promoted as a 'paradise unexplored'.

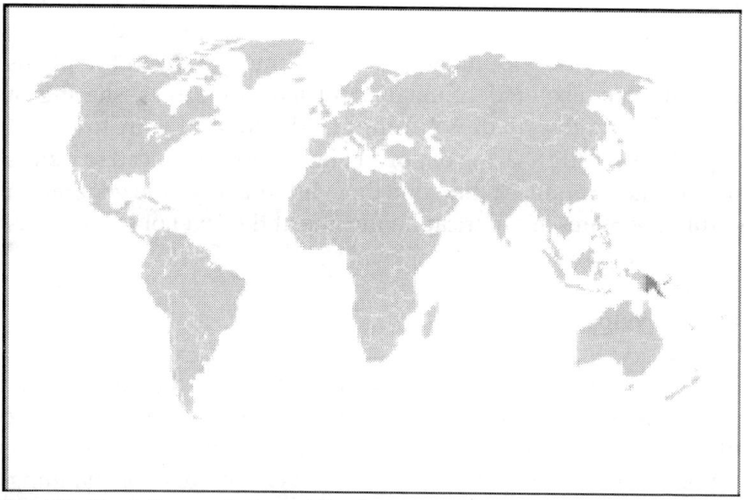

Fig. 3.2: North-Eastern India and South East Asia

The host communities who catered to the tourist's needs, attitudes, and values, tended to adapt the tourists' culture. In the process, not only the people got acculturated, but the natural environment got modified or spoiled by tourism. There has been asymmetry in host-guest relationship in the domains of culture and language acculturation, but globally, more and more communities the world over are resorting to tourism ventures especially in the developing and less developed countries.

It is incongruous that communities who take pride in their traditional cultures are compelled to take the tourism development route to community development as evidenced in the North East India context (Kulirani, 2007).

Unlocking North East India

Communities in the North East India in particular are keen to avail the tourism business opportunities. The destination *'India's North East'* is being marketed as *'Paradise Unexplored'* by the Ministry of Tourism, Government of India. The official position of the Government of India is that the land-locked North East is the gateway to South East Asia under the 'Look East Policy' to open up and promote North East to forge economic cooperation and promote tourism (Cf. Bezbaruah, 2005). 'The infinite variety, its geographic setting, its topography, its varied flora and fauna and avian life, the history of its people and the variety of its ethnic communities and their rich heritage of ancient traditions and lifestyles, its festivals and crafts make it a holiday wonderland that is just begging to be discovered afresh' as *'India's North East'* is officially described by the 'Traveller's Companion' (2005). The above portraits mostly depict the tribal life in full splendour, variety and colours.

Tourism promotion was identified as a means for development of the North East by the Government of India, and the Shukla Committee report on North East Development in 1997 in their main recommendations on tourism had suggested the establishment of a North East Tourism Development Corporation which, along with the North East Council (NEC) under the DONER Ministry and various State tourism departments would prepare plans for the development of tourist sites and related infrastructure in the region. More tourists will be encouraged to visit these States if the Restricted Area Permit (RAP) Regimes (for foreign visitors) are relaxed, and if it becomes simpler to acquire Inner Line Permits (for domestic tourists), which should be issued at all major tourist offices, railheads, airports and State entry gates, and along national highways.

Most of the State Development Reports of the NE States (Assam, Tripura, Sikkim and Manipur) 1997, recognised tourism as an important sector for the region, which need to be developed to enhance income generation. The emphasis was on eco-tourism, and other sustainable forms such as adventure, village and rural tourism. The final study report of the National Institute of Public Finance and Policy (NIPFP), New Delhi, 2008 titled *Peace, Progress and Prosperity in the North-Eastern Region: Vision 2020* (commissioned by the Ministry for the Development of North East

Region—DONER and the NEC) concluded that "the problem of the region is not merely confined to lagging development, but extends to the process of development itself. The top down development planning strategy has not involved people in designing and implementing the development strategy and not surprisingly, the relationship between public spending and service delivery outcomes has been tenuous".

The Vision Document stresses on participatory development model wherein the villagers and communities to be involved in the design, review and implementation of rural development plans through Self Help Groups and other representative groups in addition to Panchayat Raj Institutions and district council/village councils. It is paradoxical that the report goes on to state that "the focus, however, should be on high value tourism which will require close collaboration with the private sector hospitality industry, building high quality infrastructure and well targeted promotion".

Community Efforts

The business opportunity of tourism in the context of developmental aspirations of small-scale societies similar to the Scheduled Tribes in North East India having tourism generating endowments is to be positioned in a special category that draws on the essence of ecotourism, cultural/heritage/community tourism, and responsible tourism practices. The value of tourism development to the States and communities of the region is amply clear from the initiatives that are already launched at both levels. The conceptualization and implementation of any tourism project in the region must be the prerogative of the concerned communities of the region and formulated not only to preserve and protect the values and culture of the community, but actually to enhance them.

It is, therefore, necessary to bridge the gap between expectations and performance by mobilizing social energies, forging and fostering creative and collaborative partnerships with civil society, target groups, stakeholders and grassroots institutions to induct people into the planning and implementation process in a participatory mode. Fortunately, most of the indigenous communities in the North East exhibit substantial social cohesion and the communities are willing to work in teams and groups. These social dynamics can and must be harnessed to accelerate the socio-economic development of the region (Vision, 2020). The success or failure will depend on participatory planning, management and control of the development process.

There is a passing reference in the NIPFP report that the Ministry for the Development of North East Region (DONER) needs to look closely at efforts in Sikkim, Meghalaya and Nagaland in terms of 'interest in eco-tourism and model villages that are being developed also as horticultural centres. In addition, the infrastructure provision is to a certain extent decentralized as tourism is also being put under the Communitisation programme of the State' (NIPFP, 2008). The above strategy converges with the tourism ventures initiated in Mexico, New Zealand, Australia and Canada by the confederation of indigenous communities.

NIPFP report although has touched upon the need for communitisation approach has not given concrete blueprints. Such community based enterprises have good probability for success in the context of North East and the communitisation programme as envisaged and enacted by the Government of Nagaland 'Nagaland Communitisation of Public Institutions and Service (NCPIS) Act, 2002' enlarging the scope of community participation in the development and welfare programmes in areas such as education, health, power, rural water supply, rural tourism, roads, forest, sanitation, rural childcare, etc. NCPIS Act, 2002 needs a closer look for tourism development in the context of North East especially in States where the provisions of Sixth Schedule are provided for.

Conclusion

The North East Tourism Development Corporation should be developed and marketed as a unique brand that has recognized the region's distinctive character and built partnership of all aboriginal groups, cutting across their ethnic affinities that will address the specific needs of the communities in North East India. Community participation in tourism industry could lead to drastic changes in the social structure of the community in terms of emergence of elite and non-elite classes. The elite class who negotiates with the alien agencies of development is inclined to impose alien modes of production in the context where domestic production is the norm.

In the context of 'tribal/indigenous' community development in North East India, the communities can be viewed as ethnic geographic units. Macro planning for the regions of their distribution as a whole alone may not yield positive results. The ethnic groups may be feeling threatened by the domineering nature of exogenous development planning and implementation that is submerging the indigenous. The fear of marginality originates from the fear of being swallowed up by the globalizing forces. The simple subsistent economies of the aboriginal communities need to be protected against the marauding 'new economy', though in economic terms the market forces constitute the core and the weaker economies constitute the 'periphery'.

In the context of the tribal/indigenous communities of India coalition building has not emerged in political, economic, cultural and socio-religious levels as against the Scheduled Castes/Dalit communities. There is scope for the indigenous communities to work on such a coalition. Community development through tourism development could be the focal point for this coalition building.

REFERENCES

Bezbaruah, M.P. (2005). 'Prospects for tourism', in *Seminar–Gateway to the East*. Vol. 550: 32-35. New Delhi.

Bhattacharjee, J.B. (1977). *Cachar under British Rule in North East India*. New Delhi: Radiant Publishers.

Gassah, L.S. (1997). *The Autonomous District Councils*. New Delhi: Comsons Publications.

Kulirani, B. Francis (2007). Community Tourism and the Indigenous People: A Comparative Study of the Strategies of Sustainable Community Development and Tourism Development in the Indigenous People dominated Provinces in Canada and in the North East Region of India. Report submitted to SICI, New Delhi. (Unpublished).

Ministry of Tourism (2005). *Traveller's Companion–North East India*. Government of India.

National Institute of Public Finance. (2008). *North Eastern Region Vision 2020*. Ministry of DONER and North East Council, Shillong.

Prabhakara, M.S. (2004). Is North East India Landlocked? *EPW* 39(42): 4606-4611.

Registrar General of India (2004). Census of India: Map Profile 2001,

India, States and Union Territories, Office of the Registrar General of India, New Delhi.

Shimray, U.A. (2004). 'Socio-Political Unrest in the region called North-East India', *EPW* 39(42): 4637-4643.

4
Habitus and Habitats
Resource Conflict and Development Failures among the Indigenous Communities in Wayanad, Kerala

C.P. Vinod

Introduction

Wayanad holds a special place in the development history of Kerala largely for its creation of multiple trajectories of development by the settler and 'indigenous' communities even as it shares the common characteristics of State's development experiences. The region has also witnessed tremendous changes in its physical environment due to the economic development it had achieved in a relatively short span of time, although it has been proved unsustainable in the long run. Whatever may be the quality and scale of development in the political and ecological setting of Wayanad, it is also a fact that major chunk of the marginalised *adivasi* communities in the region has been kept away from accessing minimum standards of development in terms of better health and education.

This chapter argues that the development experience of various communities in Wayanad suggests that a congenial physical environment alone cannot ensure a minimum standard of development among the communities. Social practices of the communities which are shaping their habitats and are in turn reshaped by their 'habitus' (Bourdieu, 1977) are decisive in creating differential development paths and defining their successes and failures.

The term habitat, as per its dictionary meaning, is the usual natural place and conditions of growth normally referring to animals and plants. The term is used here to mean not only the physical environment of people's living, but also a social space. Communities are in constant interactions with each other, and with their physical environment in constructing a suitable social space for themselves. Social practices of the communities in creating such social space of living are derived from their 'shared dispositions' or values which is understood as *habitus*.[1]

This chapter is structured in three sections followed by a conclusion. The first section introduces the physical environment and people of Wayanad in general, with a special reference to a micro locality to empirically trace out the changes. The second section deals with the historical changes in terms of migration, interaction between communities and their habitats in 'development' and 'pre-development' scenarios. Social practices of the communities in relation to their respective habitus are traced in the third section to understand their development trajectories.

Physical Environment and People

Geographical Location

Wayanad district is situated on the northern part of the Kerala State and is the continuation of the great Mysore plateau and is contiguous to the Nilgiris of Tamil Nadu. It is bound on the North by the Coorg district of Karnataka, and on the south by the Kozhikode and Kannur districts of Kerala. It lies on the southern top of the Deccan plateau and its attraction is the majestic Western Ghats with its lofty ridges interspersed with green forest, deep valleys, and rugged terrains.

Wayanad is in an elevated picturesque mountainous plateau in the Western Ghats. It is lying between east longitude 75°47′23″ and 76°26′40″ and north latitude 11°58′40″. It covers an area of 2124.7 sq. km.

With the elevation ranging from 700 to 2100 m amsl, according to a classification (GoI, 1997) the district is divided into four geomorphologic units.

(1) *High ranges with rugged topography*: The general elevation is between 1400 and 2100m amsl and is characterised by highly rugged topography with steep slopes and narrow valleys. These hills are covered in dense forest.

(2) *High ranges with moderate rugged topography:* This includes the hill ranges along the eastern part, and the isolated hills on the rest of the area. The hill ranges occurring in the south-eastern part of the district is making a boundary line between Kerala and Tamil Nadu. The altitude ranges between 1000 and 1400 m amsl.

(3) *Mid-land region:* This is the most extensive topographical feature covering 63 per cent of the total area. The altitude is between 750 and 1000 m amsl

(4) *Inter-mountain valley and floodplains*: The inter-mountain valleys are very narrow in the high ranges and wide in the mid-land regions. They consist of both the depositional and erosional valleys. The flood plains are restricted along the river course and mainly noticed in the central and eastern parts of the area.

In this chapter we focus on a mid-land region village in Wayanad which is inhabited by both migrant population and indigenous or native communities. This village is a relatively secluded region surrounded by forest on three sides and river *Kabani* flowing through the western side of the village. This is also a typical Wayanadan village with hill, slopes, valleys and forest.

Climate

The distance from the mean sea level and the surrounding forest has given it a friendly climate. Generally, the year is divided into four seasons, viz., cold weather (December to February), hot weather ((March to May), South West Monsoons (June to September) and North East Monsoon (October to November). During the hot season, the temperature goes up to a maximum of 35°C and during the cold weather it goes down to 15°C. The average rainfall is 250 cm per year. Rainfall in the district is highly seasonal (Planning Board, 2004).

Flora and Fauna

The flora of Wayanad is typical of the Western Ghats and the plantation crops grown in the cool climate. Many of these are not indigenous to the place, but introduced in the colonial times as part of the origin and development of plantations. A major portion of the district is covered by coffee. Arecanut, palms and jack trees are also grown here. Tea is grown as an industry in large estates. Trees of the wild type like rosewood, *anjili* (*artocarpus*), *mullumurikku* (erthrina), and several species of caussia and many other non-descript varieties are still preserved here and there, to give shade to the coffee plants. Silver-oak, which is suited to the cold climate and a quickly growing tree, is planted widespread especially in the coffee plantations for shade and also for supporting pepper vines. A shorter variety of eucalyptus is also cultivated on a large scale in certain parts of the district.

Forest

Forests in Wayanad fall under three categories (i) Plateau Deciduous, (ii) Tropical evergreen, (iii) Tropical semi-evergreen. The most common is Plateau Deciduous found at about 7,00-1,000 metres above the mean sea level, mostly located on the eastern side of the district on the Begur and Chedeleth range in an area of high precipitation. This area has a valuable belt of teak forests. The tropical evergreen forests found at a height of 3000 metres and above mainly concentrated in the Lady smith Reserve forest in Chedeleth range. The semi-evergreen forests are also found here and the teak is also abundantly grown. Around 38 per cent of the total area in Wayand comes under forest. It comes to around 8 per cent of the total forest area in the State.

Cropping Pattern

Literally, Wayanad means the land of paddy fields (*Wayal*=paddy field and nad = land), which was apt till the massive conversion of paddy fields began in recent times. This high altitude district is characterised by the cultivation of perennial plantation crops and spices. The major plantation crops include coffee, tea, pepper, cardamom and rubber. Coffee-based farming system is a notable feature of Wayanad. Coffee is grown both as a pure crop and as a mixed crop along with pepper.

Here, pepper, rubber and coconuts are cultivated on the top plateau and coffee is cultivated on the slopes. Paddy, banana, coconut and arecanut are cultivated mainly in the plains area.

People

Density of population in Wayanad is 369/km^2, the least in the district. According to the 2001 census, the total population is 6,72,128 of which male population is 3, 41,958 and female 3,30,170 Adivasi population in this district is the highest among the districts in the State with 36 per cent of the total *adivasi* population in the State belonging to Wayanad district. Scheduled Caste population comes to only 1 per cent of the State population (GoK, 2003).

In the micro locality, where we try to trace out the changes empirically, there are about 683 households with an estimated population of about 3,050. Out of the total population 57 per cent are settlers, predominantly from the Christian community, followed by Ezhavas. The rest 43 per cent of the population are natives who have been living here for centuries. They belong to the communities like Paniyan, Kattunaykan, Mullukuruman and the Chetty; of these the first-three are tribal communities constituting about 32 per cent of the population. The Forward Community comes to around 28 per cent of the population of which the majority are the Christians. Likewise, more than 60 per cent of households are of settlers (Nair, Vinod and Menon, 2007).

Migration, Changing Habitats and Development

Migration and the Genesis of Development

Large-scale migration to the Western Ghats region of Malabar marks the beginning of development in its modern guise in Wayanad. Stories of development successes are often narrated here on par with the localised experiences of clearing the forest creating a standard phrase like *kaadu nadaya katha* which means the story of how the forest (*kaadu*) became mainland (*naadu*). The protagonists of development were the migrant communities who came to the upper terrains of Wayanad from the plains of Kerala in search of new lives and livelihoods.

'Malabar migration', which changed the destiny of entire Wayanad, began in the first half of the twentieth century following the Second World War. A large number of people started to move from the Southern Kerala to the Malabar district mainly due to the poverty caused by the world war and the shortage of land as a result of expansion of cash crop plantation in those areas. Europeans had occupied large tracts of lands in many parts of Travancore, with their influence in the State politics, while large number of indigenous cultivators grabbed the available government lands.

Due to the resultant shortage of land since the 1930s, people have moved to Malabar, including Wayanad, where large tracts of forests and revenue lands were left open to be encroached upon and vast fallow lands were kept accessible for rent by the landlords. That is why the Malabar migration is observed to have been the result of the explosive development of plantation crops in Travancore (Baak, 1997).

"It was the Syrian Christians belonging to the Roman Catholic denomination who were predominant among the migrants who had the initial advantage of getting information regarding the cultivation of commercial crops" (Tharakan 1976; Baak, 1997). These migrants reached Wayanad mainly in three phases starting from 1930s till the end of the 1970s. In the initial period, big farmers reached in few numbers but occupying large tracts. In the subsequent phases, Waynad witnessed massive inflows of small farmers, poor peasants, and workers.

Though migration changed the landscape of Wayanad in unprecedented ways and redrew the habitats of communities in newer patterns, it was the colonial regime that brought the seeds of transformative changes in the region much before migration. Clear felling of natural forest for constructing teak plantation was a major offshoot of colonial administrative policy grounded on the broader economic policies related to the imperial demands. Rampant commercialisation of forest resources was also a 'contribution' of colonial government. They replaced the subsistence value of forest with market values by systematising the timber trade with the introduction of chain of depots and auction centres across Wayanad. The British also introduced large scale commercial cropping of coffee from the 1840s onwards. Large areas of natural forest were converted subsequently into monoculture commercial plantations. The introduction of tea estates in the early twentieth century led to the disappearance of the forest on a massive scale. The great need of the British forest officers to earn revenue and timber from Wayanad led to over-exploitation of trees.

Migration further expanded these processes to the nooks and corners of Wayanad with the felling of forests and the expansion of cultivation of commercial crops, predominantly coffee and pepper. Migrant communities could reap maximum outputs from the fertile virgin soil of Wayanad in the later period. It also accelerated the economic development of the region subjected to the exploitative intermediary structures and fluctuating global markets.

Apart from Christians, Ezhavas, an intermediary Hindu caste, was an important migrant community in the area which established an economic base in the region through this process. Their trajectory in this development process was, largely under the impact of a reformist movement in Southern Kerala based on the teachings of Sree Narayana Guru. It urged them to move forward to find new avenues of economic mobility and they reached Wayanad either on their own or attached to their Christian friends from erstwhile Travancore.

Meanwhile, indigenous communities were largely confined to the ecological recess of wet land cultivation. Shifting cultivation which fed the local *adivasi* for centuries came to an end with no possibility of revival partially due to the government policy prohibiting it and also due to shrinkage of the area with the large-scale colonisation of plains people. Millets like *chama* and *ragi* faded out of the usual menu of the communities like Chetty, Paniyan and Kurichya.

Habitats in the 'Pre-development' Period

Closer picture of a village life shows us the meaning of changes in their habitats under specific social practices and the meaning of economic development. We can classify the communities of this village largely into two categories based on their social practices in using the physical environment. This classification is analytically used as these practices have a decisive role in determining the pathways of development for these communities.

The first is the Settler communities mainly Christian and Hindu Ezhava communities, migrated mostly from southern districts of Kerala, who introduced cash crop cultivation in the village. These communities were the agents of market based social and development practices. The second category is the 'Native communities' who traditionally depended on wet land cultivation and/or forest gathering, some who were also shifting cultivators, organised their social life through symbiotic interactions among the communities and nature.

Settler communities in this village, as already noted, are mainly those who migrated from the southern parts of Kerala since the second half of the last century with an intention to occupy fertile lands availed cheaply both in revenue and forest lands. Christians and Hindu Ezhavas are the predominant communities in the village among the settler communities. They cherished new dreams attached to these promised lands in relation to an emerging cash crop economy. Intensive dry land cultivation of cash crops became a major locus around which these communities organised their social practices.

Meanwhile, Chetty communities in the village, comprised of three different sub-groups namely Waynadan Chetty, Idanadan Chetty, and Mandatan Chetty, the prefix to Chetty signifying their original geographical locations, had created a habitat based on wet land cultivation. Physical properties of this habitat were mainly slopes of the mountains on the edges of vast wet lands which they called '*korava kandam*'. Other original inhabitants, *adivasis*, who are divided in three different groups in the village, viz., Paniyan, Kattunayakan and Kuruman, also shared the common properties of this physical environment. Spatial configuration of the habitats for each community differed based on their belief system around the places.

Belief system linked to ancestral worship, attachment to common cremation ground and yearly ritual practices, etc., had important role in constructing the idea of habitat among the adivasis. For instance, though Paniya stayed in the same patches attached to the Chetty households, their emotional bondage to the land was mainly linked through ritual practices associated with various beliefs. There was a practice of celebrating dry funeral in which they used to conduct ceremonies to please the ancestral spirits by making offerings. There were also annual offerings to their god *Gulikan* in a particular place under a big banyan tree on the common ritual ground which connected the Paniyas families in the village who are relatives by blood or marriage. Meanwhile they were not allowed to go to the worship places of Chettys. So the physical movements of the communities also were restricted in relation with their belief system. For Chettys, paddy fields were sacred one. They offered special ceremonies during the time of planting of seeds and harvest time. But Paniya workers were not directly part of these ceremonies conducted on the fields. For them paddy fields remained as a space for work and celebrating the work. Special labour arrangement like '*kamabalamnatti*' was a case in point, in which maximum number of Paniyas was brought to work accompanied by music and dance on the field.

Study locale which is referred in the chapter is comprised of four major *Kunnu* (hill) that lay parallel. A *Kunnu* (literally hillock) refers to a traditional neighbourhood comprising of generally landholding Chetty households and labouring Paniya or other *adivasi* households. The place name prefixed with a *Kunnu* is the household name of traditional Chetty joint family in the study locale.

Each Kunnu is surrounded by large areas of paddy fields. In this locality, all the traditional Chetty houses overlook paddy fields whereas houses of the settler farmers are mainly located facing towards the roads. Settler farmers, when they came as migrants, occupied mainly the garden lands, which had no importance in the Chetty scheme of agriculture.

During the pre-development period, Chetty communities and *adivasis*, especially Paniyan, nurtured a special kind of ritually based livelihood relations around production process based on wet land cultivation and cattle rearing in large numbers. Paniyas were known to be endowed with special skills for working in the marshy lands which was locally known as *korava* or *aathi kandam*. Indigenous technology coupled with Paniya expertise in toiling on the marshy land associated with a series of ritual practices molded the habitat of social interaction among the communities. Locally fixed measures of paddy played as the medium for intercommunity exchange in those days. Paddy decided the value of services and commodities exchanged. Chetty community was the centre of this transaction chain as they owned and managed the production.

Habitats in the Development Scenario

After the colonisation of the settler communities, physical characteristics of the locality were thoroughly changed with the introduction of new meanings for spatial construction. Development of the local infrastructure in terms of the construction of roads, bridges, electricity and public utility centres such as schools and health centres and proximity of the communities with it could reconfigure the spatial distribution of the communities according to their prominence.

Communities concentrated on both the sides of the roads were mainly settlers. *Adivasis* and other traditional land owning communities largely remained in their old settlements. Most of the early roads in the village were constructed by the settlers. Meanwhile in the later years, many roads were built to the tribal settlements under different government schemes which were also utilised more by the settler population as their understanding of its utility was higher.

In our study locality, both sides of the main road are occupied mostly by the well-to-do households belonging to settler communities. Usually these pieces of land are more expensive than the land in the interior village. Most of the occupants of these lands are relatively early settlers who could occupy such lands and some of the other households in the same route belong to salaried groups who bought the plots by paying cash in more recent times. Chetty and *adivasi* neighbourhoods within the forest have poor connections to these main roads.

Well-to-do households situated on the sides of the main road in the locality are seen to have built strong brick walls along the plot making a boundary to the road, but with iron entry gates. Many households in the lower economic strata use a kind of thorny shrub fence for the same purpose. No *adivasi* household in the locality is found to have put up walls, although some of them have tried to imitate the fencing even when there are no roads.

Social Practices, Changing Habitats and the Meanings of Development

Major shift in the social practices is mainly linked to the production process. Dry land became important in developing a cash crop economy. Market relations were introduced in every sphere of social life. Utilization of land became more insensitive and technology intensive. Drastic changes in land utilisation practices and large scale clearing of forest resources have done irreparable damage to the local ecology. Widespread cultivation of the crops like tapioca resulted in soil erosion. Encroachment of forest and deforestation also have made long-lasting changes in the local environment.

Ecological degradation is also observed to aggravate the economic degradation of *adivasi* life in the area, along with the destruction of common resources which existed as a life supporting system for the indigenous people. *Adivasis* can explain how the local ecology has changed over a period so intensively, which, in turn, has increased their livelihood struggles. Increase in temperature has also resulted in the reduction in the availability of tubers and roots. The area of pastureland also has been reduced considerably, which has adversely affected some of the tribal families who depend on cattle rearing as a way out of livelihood crisis and have no accessibility to the local gardens for pasturing.

Even according to the perceptions of farmers, the use of chemical fertilisers and pesticides without any discretion ended up in the outbreak of widespread pest infestations and diseases and deterioration of soil quality. Changing climatic pattern is also perceived as a reason for the change in soil quality. This is one of the main reasons for the decline in productivity of the crops like pepper in the latter years.

There is a large-scale conversion of paddy land mainly due to the intensification of commercial cropping. Permanent conversion is due to the cultivation of mainly arecanut and coconuts. Once it has been permanently converted, people also started intercropping with the cultivation of pepper and coffee.

One major impact had been the drying up of paddy fields, which has had a chain effect. If a plot is converted, it causes shortage of water compelling the neighbour to follow suit. Many people in the area observe that the lowering of water level at the water sources like ponds and wells is a consequence of conversion.

Paddy conversion also resulted in the levelling up of the traditional water pits, locally known as *'Oli'* used as water sources for domestic purposes by the traditional communities of the area, which were plenty in earlier days. Marshy paddy fields, once found commonly in the whole of Wayanad, have almost disappeared from the area. These were also the traditional water pools that absorbed the streams from the forest valleys. Factors like changes in agricultural practices and climatic conditions together have contributed to the drying up of ponds. Many such fields had been dried up artificially even in the initial phase of migration, by filling them up with the leaves of teak tree as part of their attempt to domesticate the environment and make it suitable for agricultural work.

Native communities in the village largely abstained from these practices, though in the later years these became the standard practices in the region. Still there are households belonging to native communities, though only a few, who don't use fertilizers in their fields. These are from Chetty and Kuruma communities in the village who continue paddy cultivation on a wide scale, although it does not yield much profit to them. Many Chetty households are not ready to convert the wet land simply on the ground of sentiments, not going by the economic rationale of the settler communities.

Marginalization of *adivasis* in the village was complete with the alienation of land and loss of common resources to which they had open access. Even the changes in the physical environment resulted in the distinct patterns of the habitat. They were pushed to the corners and rehabilitated in exclusive settlements called 'colonies'. For instance, shift from *paadis* (term used by the Paniya community in the village to refer to their traditional settlements) to rehabilitated tribal colonies (cluster of houses constructed under government schemes) became instrumental in the erosion of their collective dignity.

Even the emergence of strong labour market in the region did not make any significant change to the labour practices among the Paniya community which is almost a hundred per cent labour community in the village. They could not diversify their economic activities. They remained as labour bands bound mainly by kinship or friendship. Many of the Paniyans continue certain cultural traits in their behaviour carried over from their traditional reciprocal system. Many among them are accused of stealing crops from the garden of local farmers. This is true to a certain extent, as they believe that they have rights to collect food crops from the local gardens or fields, which they enjoyed for long period in the past.

In the *valli* system they had every right to do so, if their masters failed to serve required amount of food. They mainly collect fruits including mangoes, jackfruits and other tubers. Usually they do not touch cash crops.

Paniyan and Naykan women also collect the leftovers from post-harvested gardens. Women belonging to neighbouring families or same neighbourhoods move from one garden to another for collecting the abandoned remains of crops like pepper, coffee, ginger, tapioca and paddy. They can now do it only at the mercy of landowners. Some of them even let their dogs to chase these women. Some farmers were publicly seen accusing these marginalised women as thieves.

Migrant population in the village has given utmost importance for giving education to their children since early days. Modern education and health care practices had been given primary attention by the pioneers among migrants in the village since the formation of their settlements in the area. Education had caught the imagination of settler population in diversifying their livelihoods and in acquiring further socio-economic mobility. This was one of the major dispositions in the development habitus of migrant communities.

In the case of marginalised *adivasi* communities like Paniyn and Kattunaykan in the village, the situation was, however, pathetic. Most of the *adivasi* students in the village do not go to school beyond the 5th standard. Many among them are illiterates. There are a number of reasons for the irregularity in attending school by tribal students. Seasonal shift of residence between the villages of father and mother depending on the availability of work becomes a major problem in continuing education of many of the Paniyan children. Apart from the work related shifts, they also travel around the households of relatives during the festival seasons. This is common among Paniyan and Kattunaykan. Elder members, especially old women, visit the households of blood relatives occasionally. They bring their school-going grandchildren along with them and would return only after three or four days of stay. For some other students, especially girls, it is difficult to go to school during the period of harvest. When the parents go to the fields, they stay back at home to look after the younger ones. These are some of the factors that hinder the children from going to school regularly. According to teachers, they become very lazy after an interval and it will be difficult to teach them and make up for the gaps in their learning due to their absence.

Conclusion

In the case of Wayanad, a development habitus with major dispositions of market based monetary regime was organising the practices of migrant or settler communities at large. They imbibed the principles and practices of market economy at greater speeds even at the early stages of development. Social reciprocity and ethnicity remained the major dispositions of most of the indigenous communities. While their habitats underwent tremendous changes, these were unwarranted in their habitus.

Closer understanding of the changes in physical environment and habitats in this micro locality in Wayanad shows that differential development trajectories of indigenous and other communities in the region are not decided by the physical environment *per se*. Social practices of the communities is important in understanding changing habitats within the given physical environment which ultimately explain the meanings of failures and successes of development.

NOTE

1. Habitus is 'understood as a system of lasting, transposable dispositions which, integrating past experiences, functions at every moment as a matrix of perceptions, appreciations, and actions and makes possible the achievement of infinitely diversified tasks' (Bourdieu, 1977: 82-83)

REFERENCES

Baak, Paul Erik (1997). Plantation, Production and Political Power: Plantation, Production in South East India in Long Term Historical Perspective. New Delhi: Oxford University Press.

Bourdieu, P. (1977). *An Outline of a Theory of Practice*, Cambridge: Cambridge University Press.

Government of Kerala (GoK) Economic Review (*Various issues*).

Government of Kerala, State Planning Board, Thiruvananthapuram.

Nair, K.N.; Vinod C.P.; and Vineetha Menon (2007). Agrarian Distress and Livelihood Strategies: A Study in Pulppalli Panchayat,

Wayanad district, Kerala, Working Paper 396, Thiruvananthapuram: Center for Development Studies.

Tharakan, Michael, P.K. (1976). Migration of Farmers from Travancore to Malabar from 1930-60: An analysis of its Economic Causes, Unpublished M.Phil. thesis, Thiruvananthapuram: Center for Development Studies.

5

Assimilation Process of the Tea Community in the Socio-Cultural Life of Assam

Breach and Prospects

Amalmani Sarma

Introduction

Tea, the most ancient and popular non-intoxicating beverage around the world, is said to have had its genesis in China. The Chinese believe the beverage to be discovered by their mythical emperor Shen Nung in 2732 B.C. According to *Encyclopedia Britannica*, tea originated in China about 700 B.C. Derived from the Chinese Fukien dialect, the word *Cha* is used by the Chinese to mean tea; in Assam, the beverage is known as *Chah*, pronunciation being same as that of the Chinese *Cha*. Tea is a favourite drink for almost half of the total population of the present day world.

Tea is an integral part in the day-to-day life of the Indian people. Tea is such an element of the Assamese culture without which the Assamese feel incomplete in their usual hospitality. Assam produces 53.16 per cent of the total tea production in the country and 55 per cent of the total global production (Kurmi, 2007). Assam tea has become a global brand with its quality product; it has earned glory to be one of the best in the world in its taste. Tea companies like Williamson Magor, Tata, Birla, McMillan, Khaitan, HUL, Everyday Group, etc., have been making the tea of Assam a household name throughout the country and the world for its taste. Besides the one-horned Rhinoceros found in Kaziranga National Park, the World Heritage Site of Assam, the State has occupied a place of pride in the world atlas because of its quality tea with great taste which has been included as a special item in the menu of restaurants in different corners of the world. It may be noted here that a particular brand of tea from Assam (Parry Agro, Dhekiajuli) found its way to a scene in the famous Hollywood movie 'Stuart little', reminding us about the widespread popularity of Assam tea. The tea industry of Assam may be considered as a black diamond that has become a great source of foreign exchange earning Rs. 1028 crore in 1990 (Kurmi, 2008). In a recent context, there may be a different picture as given in a report on February 29th, 2009 in *The Assam Tribune*, a leading English daily of Assam, that the global meltdown may lead to drop in the export of tea from India to countries like USA, UK and Russia.

However, this chapter is not going to directly focus on the tea of Assam or its quality and commercial aspects, etc., but on the people who have been rendering their toil and services under the most trying circumstances that has lent Assam the name she has earned for her tea. Tea plantation is an agro-based labour intensive industry. It requires a large number of hard working labourers to engage constantly in the tea gardens. Mentioned here as 'Tea Community', these groups of people were imported to Assam from various corners of India by the British planters during the colonial period to engage as labourers in the newly established tea gardens of Assam. They have been known by different nomenclatures like 'Coolie', 'Bongali', 'Labour', 'Majdoor', 'Bagania', '*Adivasi*', etc. some of which are quite humiliating for the community as the uses of such words clearly indicate a feeling of negligence towards them, treating them as 'sub-human beings'. Sarma (2006) has used the term 'Jharkhandi Adivasi' to refer to the community. Literally the term *adivasi* denotes aboriginals or aboriginal settlers. These people who belong to Austric linguistic group can be called *adivasi* in the context of the places from where they were brought, but not in the context of Assam. Tribes like Bodo, Mising, Karbi or Rabha, etc., are the *adivasi* or aboriginal people of Assam. Again, not all the people of the Tea community were from Jharkhand alone; different subgroups from different corners of India formed the large Tea Community of Assam with their amalgamation into a single entity. Moreover, while they have accepted Assam as their motherland, why should there be a label on them as 'Jharkhandi'?

In this chapter, they have been addressed as 'Tea Community' to show them as the people closely related to the tea industry in Assam. As Assamese people are proud of their tea, the term of address as the 'Tea Community' of Assam, may enhance their status in the eyes of Assamese people. The term 'Tea Community' might and (possibly) should be considered as an honour rather than a humiliation to them.

Historical Background

As mentioned earlier, tea is believed to have originated in China with planned commercial cultivation there. It would be relevant here to study the context that led to the genesis of tea industry in India, particularly in Assam. Assam went under the British rule in 1826 by the signing of Treaty of Yandaboo and became a part of Colonial India.

The tea-addicted Britons used to import tea from China for their mass consumption. But at a time when the trade relation between the two nations began to decline, it made the English to think of a new alternative for their much needed import of tea. They were determined to destroy the monopoly of the Chinese in the tea industry and became successful in their design by discovering tea and establishing tea industry in Assam during the 40's of the nineteenth century. Charles Alexander Bruce is called the 'Father of Indian Tea Industry' who was the first to make the British authority know about the existence of indigenous tea plants in Assam. He contacted Maniram Dutta Barua, popularly known as Maniram Dewan, the first Assamese tea planter in January 1825. This wealthy Assamese nobleman was the first to bring the indigenous Assam tea to the notice of the British. He introduced Bruce to a Singpho (an indigenous tribe of the region) Chief with whom his brother Robert Bruce had contracted to supply tea plants and seeds during his next visit (Singh *et al.*, 2006). With the initiative of Bruce the first consignment of Assam tea was sent to England in 1838, and the first sale in Calcutta held in 1841. The competitive quality of Assam tea was finally established in the global market in 1852 (*ibid.*).

After the discovery of tea in Assam, the British planters evinced their keen interest in the tea industry. The marvellous richness and feasibilities of the soil and the favourite climate of

Assam for tea cultivation encouraged the British enormously to acquire the virgin territory of Assam for tea plantation. Accordingly the land was cleared; the steamy verdant hills were converted into rich tea plantations to make it a booming business in the near future. The highly labour intensive tea plantation required a large number of people as workers. Though initially it was expected to be arranged locally, subsequently it became a great hurdle to have sufficient number of workers for the tea gardens. It was because of the locally unavailable workers that led to the drafting of people particularly from the overcrowded and economically depressed areas of East India to work in these estates. Further the British investors and entrepreneurs considered Assam to be safe, compared to the disturbed provinces of India during mutiny, and preferred to invest their money in the tea plantation of Assam (*ibid.*).

In the initial stage, the tea industry in Assam began to suffer because of the scarcity of sufficient number of local workers. By the time, Assam had been greatly depopulated by diseases, civil conflict and the Burmese invasions (Weiner, 1978). However, the British tea planters opened their tea gardens by recruiting the local Assamese people as tea garden labours. But mainly because of three reasons (Singh *et al.*, 2006; Baruah, 1994) the Assamese people were reluctant to work in the plantation industry :

(i) Assamese people were lazy and indolent and the workers were addicted to opium;
(ii) Assamese people were comparatively prosperous with much land in their possession and were engaged in self-sufficient agrarian economy; and
(iii) Their 'ego problem' based on their socio-cultural background restrained them from working as tea garden workers under the alien tea planters.

At that point of time, their demands in terms of food, clothing and other material aspirations were confined to bare minimum: their food consumption included to a maximum of two to three items like rice, '*khar*' (soda or alkali), etc. They never used to import any food or cloth items from outside except for salt.

A large number of Chinese labourers were also recruited in

Assam to engage in the tea gardens. But it proved futile because of their aggressive nature and quarrelled with the natives or *vice versa*. Moreover, they did not listen to their authority and had to be paid high salaries. They were mostly employed in tea factories for manufacturing and technical works. The attempts to recruit the indigenous tribal people like Singphos, Kacharies and Nagas also failed because of similar reasons (Baruah, 1994; Kurmi, 2007).

This made the British planters import workers from outside the State, especially from the thickly populated neighbouring provinces of East India (Singh *et al.*, 2006). They were mostly drawn from the tribal belts of Chhotanagpur, Orissa, Madhya Pradesh and present day Andhra Pradesh. Some other workers were brought to Assam from Bihar and the eastern part of U.P. The British knew the tribal people of Chhotanagpur and the plains people of Bihar and U.P. very well, who showed their hard working nature in the sugar plantation in Mauritius and the West Indies and accordingly these groups of people were preferred by them to be imported into Assam as tea garden labour. Moreover, the British government was giving a helping hand to their tea planters who became interested to invest their money in the less disturbed Assam after the revolt known as Sepoy Mutiny, or from another view-point, as the First War of Independence in 1857. These people became the victims of the planters and local *zamindars* in parts of eastern India. Disturbed situation in those areas led to migration of these people to Assam (*ibid.*)· The Santhal Uprising (1855-57) which extended from Bhagalpur in the northern part of Bihar to Orissa in the south stretching from Hazaribagh to the borders of Bengal; the Indigo Rebellion (1859-60) that was originated in Govindapur village in Nadia district of Bengal and the Munda Rebellion (1899-1900) in the south of Ranchi created severe administrative disturbances for the British Government. The importation of these groups of people also could be helpful for the British administrator to curb the tribal uprisings. Besides, they would be unable to rebel against the British Government in a new place as they would have to face the hostility from the natives in an unfamiliar environment[1].

Assimilation

With the frequent occurrence of calamities such as drought and flood accompanied with epidemics, poor and innocent people were dying with hunger in certain areas of Eastern India such as Chhotanagpur, Santhal Pargana and some tribal belts in the region. The uprisings caused great turmoil and led to the destruction of men and crops adding to the miseries of the local people. The recruiters took undue advantage of the situation and their helplessness that made them to migrate to Assam as tea garden labourers. These easy going ordinary workers were brainwashed and were promised "easy work, better pay and ideal condition of work and unlimited land available for cultivation" (Singh *et al.*, 2006). However, a large number of them died on the way to Assam, most of them afflicted with cholera that was rampant, and few escaped the fever (*ibid.*). The flow of these workers to Assam has continued and they have remained neglected and oppressed more by their masters than they experienced earlier.

Along with their own culture they have been influenced by certain elements of the culture of their adopted land. In course of time, they began to be assimilated in the socio-cultural environment of Assam with their own distinctive culture. With their assimilation, a common identity has been formed among the Assamese people including various other groups under a larger framework of Assamese lineage. It has strengthened the process of formation of the greater Assamese community.

There are more than eighty sub-groups within the Tea Community in Assam (Baruah, 2008). Munda, Santhal, Orang, Bheel, Telengana, etc., are some of such groups who had diverse traits and attributes of culture and dialect as they hailed from different parts of the country. Having been influenced by the local culture and the culture of one another they have assimilated to be one single entity and has enriched the Assamese culture. Most of the people brought to Assam as tea garden workers did not go back to their places of origin after the completion of their contracted term and settled in the State permanently. Some others began to be settled outside the confines of the tea plantations and scattered in the nearby villages of the tea gardens (Deka, 2008). The people of the tea community have never been considered to be a cultural, political or economic threat to Assam by the rest of the population; it is because the jobs that they held never attracted the attention of local Assamese people. They mostly made them confined within the territorial boundaries of the tea gardens and never have raised their voice against the interests of Assam and its people. They have never shown their interests to encroach upon the lands of local Assamese people (Kashyap, 2008). This considerable amount of assimilation can be better understood in the light of the influences and exchanges in different aspects of the society during the long period of more than one hundred and fifty years.

Despite the assimilation and non-challenging attitude of the people of the community, they have retained certain distinctive traditions of their own in a splendid way. Festivals like *Tusu, Karam,* etc., are celebrated by them with great enthusiasm. *Jhumur* dance is a unique element of the specific culture of the community.

This tendency of the people of Tea Community to accept Assamese as their own language helped them to assimilate with the Assamese society and culture to a considerable extent. Majority of the children from the Tea Community go to Assamese medium schools. Though both Hindu and Christian people are there in the community but almost all of them observe Assamese Hindu rituals; *Bihu* is quite popular among them. Holi is also observed by the people of Tea Community.

There are two categories of people in the Tea Community in terms of their jobs: TGL (Tea Garden Labour) and Ex-TGL. The former is a full time worker in the tea garden and the latter is seasonal one. People who belong to Ex-TGL have now been engaged in different other commercial activities such as paddy cultivation, small businesses, etc. Some of them have moved out from the territorial boundary and have begun to settle down in the neighbouring village of the gardens. The overall picture of the people of Tea Community shows them to be in a dilapidated condition. But out of this context a small number of people have established themselves in the arena of literature, education, politics etc.

Breach

However, the process of assimilation of the Tea Community into the 'mainstream' of the Assamese society cannot be said to be complete yet; it is still in the process. Majority of the population from the community simply have remained confined within the vicinities of the tea gardens. The local Assamese people, particularly the educated elite have a greater role to play to fulfil the socio-political aspiration of the community. It is a matter of fact that the people of the Tea Community are not given a prestigious position in the society by most of the people in Assam. They are considered to be the people at the lower strata of the society. They are not given an equal status with the rest of the Assamese people. Despite their imposed isolation, the Tea Community made efforts to break the confinement by starting certain strikes from the second decade of nineteenth century by acting as a united body beyond the ethnic barriers within the community. They could meet their economic demands like wage and bonus enhancement, etc., if they were properly organized; but they failed in their ends as the leading political outfits in the State from time to time have not come forward to organize them (Deka, 2008). The people of Tea Community are devoid of the basic amenities such as schooling, healthcare, reasonable living conditions, etc. They are economically stagnant, educationally backward and physically malnourished. Except for a small number of people who have come out of this situation, majority of them have to live in an unhygienic deplorable condition. Debarred from basic minimum needs, the children are either engaged in the tea gardens as child labour or employed as domestic help. Some of them are also engaged in brick industry. According to an estimate disclosed by the 'Institute for Plantation, Agricultural and Rural Works' of North Bengal, there were 41,499 child workers in the tea gardens of Assam in 1991 (Tanti, 2007). Ironically being employed in an industry which is earning a considerable amount of foreign exchange for India, the tea garden workers are living a harsh life of the poorest of the poor in the country.

The Tea Community makes up almost 20 per cent of the total population of Assam (Kar, 1990). There are almost 70 lakh of people from this community in the State who are scattered in 850 tea gardens (Baruah, 2009; Duarah, 1996) in different corners of Assam. The literacy rate of such a large group of people is so poor that it cannot leave space to think of an all-round development of the community until and unless necessary measures are taken to provide them the basic minimum educational facilities. Though they have acquired the Assamese language by adaptation, they could not go forward on an equal footing with the rest of the Assamese people which created a huge gap in the assimilation process. The people of this community have to spend most of their time thinking of their survival needs like food and clothing. Frustrated and desperate about their hard and monotonous day-to-day life and the uncertain future, most of the workers have become habituated to intoxicants, particularly liquor. Significantly there are hardly any students from the Tea Community who have got the opportunity to get education in 'Assam Valley School', one of the most sophisticated and costly schools in Assam founded by Williamson Magor, a mega tea company of the State which would not have become what it is today but for the hardship and toil of the tea workers of Assam.

A majority of the hierarchy of Assamese officials in the tea gardens still like to maintain a medieval master-servant relationship with the workers which is humiliating not only for the latter but for the entire Assamese community. The tea garden authorities are not at all concerned with the interests of the people of the Tea Community in terms of the basic minimum needs of the day-to-day life. This is not the case only with tea companies; the State government has also not undertaken any steps for their social and economic development. The same condition of the Tea Community prevails in the tea gardens which are not owned by the Companies, but by Assam Tea Corporation (ATC) under the Assam Government. The conditions of tea gardens owned by ATC are even worse than that of the rest of the tea gardens in Assam. Although placed first in tea producing in the country, the tea workers in Assam are paid less salary than the tea workers in the States like West Bengal, Tamil Nadu, Kerala and Karnataka, etc. (Kurmi, 2007).

A large number of Tea workers are dying in unhygienic conditions in the 'Lines' or '*Basti*', the residential area meant for them; they suffer from diseases such as malaria, gastroenteritis, and sometimes, Japanese encephalitis, etc., and die in large numbers. More than 400 tea workers died in 1998 suffering from gastroenteritis in different tea gardens of the Upper Assam. Lack of proper health care sometimes makes them to suffer from deadly diseases and loss of lives. Moreover, 9 workers died out of hunger in 1994 in Karimganj districts of Assam (Kurmi, 2007; 2008). Neither the tea companies nor the State government seems to be seriously concerned about this deplorable condition of the community. The rest of the Assamese population is also unconcerned about improving the quality of life of this exploited group of people.

The social position of the Tea Community has not been

recognized till today. They are the most oppressed and exploited group of people in the Assamese society. They have been still put into a situation to be dependent on others for their socio-economic development. The Ex-TGL who are the seasonal workers of the tea gardens and scattered in the vicinities of the tea gardens in other than tea garden activities have not also been able to be self-dependent. Some of the students' organizations are demanding Schedule Tribes (Plains) status for the Tea Community as a means to rescue them from the age long oppressions that they have been suffering from and to go forward a step to compete and adjust with the present day situation. However, the educated elite Assamese have not been making them a part to the expected level in this endeavour to lead the community towards the path of socio-economic development.

Certain instances of isolated and unorganized protests of the Tea Community are frequently witnessed regarding the bonus issue and issues of their lack of civic amenities which sometimes took serious turns with loss of lives. On May 30, 2003, the workers of the tea gardens of Dhekiajuli region in the State, agitated over the issue of human subsistence, set on fire the Deputy Manager and the Field Assistant. In another tea garden, the Assistant Manager was lynched by a group of workers for dismissing them as they went on leave without any notice (Kakoti, 2007).

A Case Study: The Infamous Beltola Incident

On November 24, 2007, Assam witnessed the most shameful incident in the recent times. A procession was taken out by the All Adivasi Students' Association of Assam (AASA) participated by both male and female to add their voices to the demands for the inclusion of 'Tea Tribes' and 'Santhal *adivasis*' in the list of Scheduled Tribes (Plains). The demonstrators were to march from Beltola to Dispur in the Guwahati City to raise their voice in the State Secretariat. The procession suddenly turned violent while the police tried to stop them and some of the participants started acts of vandalism–damaging the roadside cars, shops and properties in Beltola area which resulted in brutal retaliation from the local residents. Along with the death of at least one person and hundreds injured, a young class X girl was chased by some unruly youths and stripped naked in broad day light and in full view of the public. Several demonstrators were mercilessly beaten up by some youths. All this took place within a few hours of chaos. Finally a local middle-aged person came to her rescue and several other people gave shelter to the victim. Later newspapers and electronic media published the incident, some of them with the photograph of the victim girl (*Frontline*, 2007).

This incident raises several questions including the very serious one as to whether this indicates the beginning of a mass disturbance with severe consequences in the near future.

Concluding Remarks

Despite the Tea Community's enormous contribution to the economy of Assam and in other spheres to the greater Assamese community, the community has been lagging behind in almost every aspect of life compared to the other sections in the Assamese society. The rest of the Assamese people especially the influential educated elite have a great role to play and should be adequately concerned about these victimized and abused people, to rescue them from the inhuman acute social illness they have been suffering from. The upliftment of this most backward segment of the population of Assam is the need of the hour. By understanding the roots of their age long sufferings, the problems of the Tea Community could be brought to the limelight to make the concerned sections of the society think about them.

The Tea Communities should be provided with the civil amenities to the maximum possible extent. Schooling, health care, good water, better living conditions should be made available in the 'lines' and in the vicinities of the tea gardens where people from TGL and Ex-TGL live in an unhygienic and downtrodden environment. Necessary measures are needed to be taken for the upliftment of the quality of life of these people. It is to be seriously considered why demands are being raised to include them in the ST(P) list and how they can be brought on par with the privileged sections of the society. The overwhelming sense of outrage among these disgruntled people is leading to the emergence of extremist elements within the community. Unattended to, this could have serious repercussions and lead to escalating conflicts.

NOTES

1.It may be relevant here to mention that there is another group of people from the same regions of the country with similar cultural traits who started their migration to Assam from 1881 onwards, but have not been at all in connection with the tea industry of Assam. However, the content of this chapter is not incorporated with the matter of this particular group of people. But, while discussing the Tea Community of Assam, the people of this group cannot be excluded as they all together may be said to be in the same vicious circle in terms of the socio-cultural scenario in the present day Assam.

REFERENCES

Baruah, Anjan, (1994). *Chah Shilpar Katha Aru Axom*, Axom Sahitya Sabha.

Barua, Gautam (2009). Rajnitir Merpakor Pora AATSA Mukto Hobo Paribone? *Amar Axom*, 19th February Issue.

Baruah, Pradeep (2008). *The Tea Industry of Assam: Origin and Development*, EBH Publications, Guwahati.

Deka, Kankan (2008). Colonialism, Migration and the Problematics of the Making of an Identity: Tea Tribes and Adivasis of Assam, *Souvenir of North East India History Association*.

Duarah, Chandan Kumar (1996). Axomor Chah Bonuar Durowostha aru Chah Arthoniti, *Prantik*, 1-15 October.

Frontline, December 21, 2007.

Kakoti, Manasjyoti, (2007). Plight of the Tea Labour in the Backdrop of Globalisation: A Case Study of the Sapoi Tea Estate, *Changing Patterns of Society and Economy–19th and 20th century*, Department of History, Gauhati University.

Kar, R.K. (1990). Tea Labour in Assam: A Note on the Problem of Identity, in *Tea Garden Labourers of Northeast India: A Multidimensional Study on the Adivasis of the Tea Gardens of North East India*, Ed. By S. Karotemprel and B. Dutta Roy, Shillong: Vendrame Institute.

Kashyap, Mohorshil, (2008). A Bud Two Leaves: Tea Community in Assam, *Cultural Heritage of Assam*, Directorate of Information and Public Relations, Assam, Dispur. March Issue.

Kurmi, Sushil (2007). *Axomar Chah Shramikar Avadan*, Assam Prakashan Parishad.

Kurmi, Sushil (2008). *Chah Gachar Maje Maje*, Kiran Prakashan (On behalf of Assam Sahitya Sabha).

Sharma, Debabrat (2006) *Axomia Jati Gathan Prokriya Aru Jatiya*

Janagosthigata Anusthansamuh, Ekalavya Prakashan.
Singh, S.N.; Amarendra Narain; and Purenendu Kumar, (2006). *Socio-Economic and Political Problems of Tea Garden Workers–A Study of Assam*, Mittal Publication, New Delhi.
Tanti, Kamal Kumar (2007). *Nimnabarga Samaj Oitijya*, Aakhar Prakash.
The Assam Tribune, February 12, 2009.
Weiner, Myron (1978). *Sons of the Soil: Migration and Ethnic Conflict in India*, Princeton: Princeton University Press.

6

Kadar's Tryst with Eco-Tourism at Athirapally and Vazhachal

Emerging Resettlement Issues of a Particularly Vulnerable Tribe in the wake of AHEP

R. Rajendra Prasad

The word *Kadan* in Malayalam means the dweller in a forest (plural—*Kadar*). The tribe obviously got the name because of their exclusive forest habitat. There is a similarly named community in Wayanad district also, but the two communities are separated by several kilometres of geographical distance, live in different habitats and have nothing in common. As these are also one among many forest-dwelling groups, the designation Kadan (and its plural form: Kadar) is used by the plains people to denote their habitation inside forest, but such reference is not made about other forest dwelling groups (Thurston, 1909). Many still hold the Kadar of Cochin one of the least advanced of the South Indian forest tribes. Kadar is one of the Particularly Vulnerable Tribal Groups of Kerala as per the State classification based on the criteria of pre-agricultural stage of development, low level of technology, diminishing/dwindling population, and very low literacy rates (Kakkoth, 2005).

This chapter attempts to analyze the plight of the Kadar in the core area of their habitation in Athirpally and Vazhachal in the wake of the emergent resettlement issue due to the proposed AHEP (Athriapally Hydro-Electric Project) that threatens the very survival of a large proportion of the community once the project becomes a reality. The Kadar families in the project area for all purposes would like to retain their traditional habitat.

Geographical Distribution and Population of the Kadar in Kerala

In Kerala, the Kadar population is distributed in the Parambikulam, Kuriakutty, Nelliampathy and Kodassery forest areas coming under Chittoor taluk of Palakkad district and Vazhachal, Poringalkuthu and Sholayar forest areas of Mukundapuram taluk of Thrissur district. They are also found in the adjoining Anamalai hills of Coimbatore district of Tamil Nadu. Population of Kadar in Kerala, according to the 1981 census was 1,120 and that of 1991 Census was 2021 and in the 2001 Census, the Kadar population was 2,145. The socio-economic survey of 1996-97 has estimated 1,472 Kadar population, consisting of 744 males and 728 females; and the survey conducted by KIRTADS for the action plan estimated 460 families in 1999. Their language is close to Dravidian language, Tamil, and is locally known as Kadar *bhasha* which has no script (Singh, 1994:450).

The Indigeneity and Marginal Position of the Kadar

The term 'indigenous' has to be examined in different perspectives. In a historical sense, 'indigenous' means earliest inhabitants if not autochthons. While in a relational sense it indicates people who were once under colonial dispensation and currently occupying almost without exception a poor and marginalized position in national societies (ILO Convention 169 of 1989). In the juridical sense, the term is used in respect of peoples having distinct cultures and who are under the pull of domination of others and are in the process of being dispossessed of their traditional life support resources and are being marginalized in one form or the other. It is usually equated with the tribals but considering historical connections and rootedness in the surroundings it also incorporates other social categories, specially the folk and marginal communities, the marginality of a group examined in the context of proximity or distance from the core elements of a society or culture. In short, the term indigenous may be addressed to a native population, an ethnic section of some large population, a depopulating group, and a social/numerical/linguistic/religious minority group and a marginalized segment of the society.

The word indigenous also should be used not in a chronological sense, but in the normative sense to cover people who feel rootedness in their surroundings, entertain a custodial sense about their territory and resources, bound together primarily through moral bindings and entertain a sense of reciprocity and mutuality reinforced through equalitarian ethos (Burman, 2002).

The Kadar Literacy rates have been dismally low in spite of the opening up of schools in their habitat. The dropout rates have been very high. One can hardly find them reach the higher secondary level. According to the 1991 census the total literacy rate among the Kadar was 47.5 per cent, male literacy rate being 50.5 per cent and the female literacy being 44.4 per cent. And the sex ratio of the Kadar is an alarming figure at 977 females for 1000 males. The relatively isolated nature of their habitat makes them inaccessible to the outside world in spite of modern means of communication and transportation. The few schools found in their settlements have not been able to deliver quality education to enthuse the Kadars to send their children regularly to school. Lack of motivation, and economic backwardness and low level of general awareness have contributed to the low literacy levels of the Kadar.

The marginality of the Kadar can be understood from the above factors. They also lack a proper forum and enlightened leadership to represent their causes in a just manner. The outside agencies working among them are concentrating on the issues of resettlement and relocation. Here again the Kadar is at the receiving end due to the development intervention for the benefit of the general society at large and not aimed at the welfare of the Kadar.

Dwindling Forest Resources and Changing Economy

Traditionally the Kadar has been a hunting and food gathering tribe, foraging in the forests and engaged in the collection of MFP (Minor Forest Produce) in exchange for essential items. The Kadar of Vazhachal combines NTFP (Non-Timber Forest Produce) along with hunting and food gathering. Their dependence on the forest for their food has declined over the years as they get rations from the government ration shops. Also the collection of forest resources is subject to the restrictions and conditions laid down by the Forest

Department regarding the items and quantity to be collected by each individual family. The forest resources over the years has been steeply declining due to non-scientific ways of extraction and over exploitation of the resources due to increasing pressure of both the tribal and non-tribal population on the satiable resources of the forest. It is also reported that the non-tribal persons other than the Kadar and other tribes residing in the forest area are making inroads into the forest to exploit the resources, and enticing the Kadar to collect those items not allowed by the Forest Department to extract them. The Kadar of late has taken up horticulture to augment their diet supply. They have taken up cultivation of tapioca, sweet potato, vegetables and cultivation of coconut and arecanut in and around their dwelling places. This shift has given the Kadar a sense of belonging and rootedness, compared to their earlier stage of hunting and foraging in the forests. The employment opportunities generated by the Forest Department through the institution of *Vana Samrakshana Samithi* (VSS) and its involvement in the afforestation programmes help the Kadar in earning additional income apart from their activity of NTFP (Non-Timber Forest Produce) collection for the Girijan Co-operative Society. Sedantariness is a changed economic position of the Kadar.

Eco-tourism and the Kadar

The Kadar have been living in the pristine forests after their resettlement from Ittani in the 1960s to their present abode in Vazhachal, which of late has emerged as a tourist spot with the presence of a waterfall in Athirapally and the Charpa rapids on the way to Vazhachal. The river Chalakudy can be accessed from the settlement of the Kadar, which attracts a large number of domestic tourists during the peak monsoon season from May to September. The river in full spate provides a breathtaking view from close quarters. Out of the 52 Kadar families residing in Vazhachal one member from each family is provided a job to augment their family incomes. The jobs usually include that of tourist guides and men manning the check-posts and women having assorted duties of weeding and cleaning the area in and around the riverbank. The employment in itself has brought some changes in the socio-cultural outlook among the Kadar who are exposed to the outside world through the tourism activities of the place. A part of the income generated through the tourism activities is set apart for the welfare of the Kadar. The living conditions in Vazhachal among the Kadar are better compared to that of the Kadar of other settlements who remain isolated and insulated from the outside world. The Kadar of Peringalkuthu, another tourist attraction, is not involved in any of the tourism related activities, though they are settled very much near the tourist spot.

Involvement of the Kadar in the Eco-tourism Activities in Vazhachal

Athirapally and Vazhachal Eco-tourism were started in the year 1998. The revenue generated out of the tourism related activities was exclusively controlled by the State Forest Department. The management was taken over by Vana Samrakshana Samithi (VSS) after its formation in the year 2001. The VSS employs 26 Kadars on daily wage basis in various tourism related activities as security guards, trekking guides, check-post guards, gardeners and cleaners, etc. Rs. 135 is paid as daily wage labour on a weekly basis payment out of the income generated. While in the year 2007-08, there were 7,844 labour days, (a labour day equalling labour for twenty-six people going by the number of people employed per day), and generated an income of Rs. 61,10,824, and in the year 2008-09, it was Rs. 6,480, generating Rs. 52,64,116; the amount accrued from vehicle parking charges also increased from Rs. 3,39,459 in 2007-08 to Rs. 5,62,687 in 2008-09. This clearly shows that eco-tourism is flourishing in Athirapally and Vazhachal. Various tourism related activities are efficiently managed by the VSS. Involvement of the Kadars in these activities has improved their economic position due to the additional source of income generated, apart from their traditional occupation of NTFP collection. Possibilities can be explored for developing eco-tourism in the nearby settlements of the Kadar.

Athirapally Hydro Electric Project (AHEP) and the Threat of Resettlement and Relocation

The State of Kerala has been experiencing power shortages for the last so many years and efforts are underway to utilize the few existing water resources to augment the shortage by a few mega watts wherever it is possible to tap the resources. Irrespective of the local issues that are involved related to the livelihood of the local tribal population, who are the actual inheritors and the custodian of the water resources, the proposed Athirapally Hydro Electric Project of the Kerala State Electricity Board (KSEB) which is expected to generate 163 MW of power is demanded by a large chunk of mainstream population even at the expense of displacing around fifty families of the Kadar. If the project is allowed to become a reality, besides the consequent loss of biodiversity of the area, it will also endanger the lives of the local tribals.

Impact on the Twin Waterfalls

The Athripally dam proposed to be located a mere 400 metres upstream of the Vazhachal rapids will eventually kill both the Vazhachal rapids and the famed Athirapally falls, located about 5 km downstream. This implies that, out of the total water flow in the river, more than 80 per cent will be diverted away from the Athirapally falls and will join the river about 1.5 km downstream of the Athirapally falls. The consequent, slender, ribbon-thin waterfall will only destroy the magnificent falls and drive away the tourists, thereby paving the way for the destruction of the local tourism driven economy. The twin waterfalls are a local attraction, which attract hundreds of tourists both locally and also nationally. The wildlife sanctuary nearby which is the abode of the Kadars also attracts quite a number of tourists into the heart of their territory. The Peringalkuthu dam, which is just 3 km from Vazhachal, has another settlement of the Kadar, the Pokalappara tribal colony. The dam has got immense potential for developing eco-tourism by involving the Kadars, who at present are not involved in any of the tourism related activities of the place. A carefully planned policy can turn the twin waterfalls into an ideal eco-tourism spot with the active involvement of the local indigenous Kadar.

Impact on Tribal Population

Kerala State Electricity Board (KSEB) holds that there are no tribal families in the project area. Anyone who has visited the project area could see the Pokalappara Tribal Settlement, which has 22 tribal families. And tragically, these are families who have undergone dislocation twice earlier, on account of the construction of dams upstream. It can be observed even by a casual visitor that the Full Reservoir Level (FRL) of the Athirapally dam will submerge most of their colony. Another disturbing fact is that the project authorities have not chosen to look at the existence of the Vazhachal Tribal Settlement with its 52 odd tribal families, tribal school and hostel. Although technically, this area may not fall within the project area, the fact is that it is merely 400 metres away from the dam site; it will be practically impossible for these families to continue their stay in the present location when the project work begins and more than 2,500 odd workers make their presence felt in the area. However, the KSEB and the Government have chosen not to look into the matter. In fact, the Environmental Impact Assessment conducted by WAPCOS (Water and Power Consultancy Services), a Government of India undertaking based at Gurgaon, Haryana does not even make a mention of this fact. And the study of Tropical Botanic Garden and Research Institute (TBGRI) puts the distance between the colony and the dam site at 4 km.

Ecological Damages

Chalakudy River, especially in the Vazhachal Forest Division, within which the proposed Athirapally project area falls is unique in many respects for long term conservation as pointed out by Kerala State Bio-diversity Board. Further the vegetation (special vegetation along the river) offers uniqueness unparalleled anywhere else in the State or even in the country. The proposed project would cause loss of several endemic species, loss of fish diversity and connectivity, unique area for bird conservation, loss of overall bio-diversity and loss of ecological functions and characters of the river system.

Discussion and Conclusion

The above observations make it quite clear that allowing the AHEP to become a reality would deprive the Kadar, the right to their livelihood, and also would dislocate them from their existing habitat. Loss of livelihood and habitat, and the consequent ecological damage would threaten the very survival of the community in the course of time. The attention of the administrators and policy planners and the wider intelligentsia is drawn here whether to allow the dislocation of a 'primitive tribal group' for the sake of a hydroelectric project to augment a few Mega Watts of electricity.

Efforts are underway to stop the construction of the proposed dam very much near the settlement of the Kadar in Vazhachal. The Chalakudy River Protection Forum is spearheading the agitation, which is becoming a Statewide movement with the involvement of the Youth Brigade. Two public interest litigations are pending in the High Court of Kerala.

In conclusion, it may be pointed out that indigenous communities should be made the rightful owners of their resources and should have the *de facto* decision making powers regarding the utilization of these resources so that they can be saved from marginalization by the wider society. The eco-tourism experiences the world over especially those related with the indigenous communities have shown that where the people are given a chance to participate in the management of their own resources, it has resulted in economically profitable ventures that contribute to improving their living conditions. The Kadar has to be trained to handle the various eco-tourism activities themselves. It is the opportune time to create a higher level of awareness among the Kadar and to instil in them the required confidence to manage their own affairs. In the present juncture it may seem a far cry, but efforts can be channelised in this direction to get the desired results, and at the same time, the Kadar may be allowed to retain their culture and identity and the uniquely pristine ecology they inherit.

REFERENCES

Burman, Roy (2002). Report of the ILO Convention 169 of 1989.

Edgar, Thurston (1909. Reprint 1975). Castes and Tribes of Southern India, New Delhi: Cosmo Publications.

Kakkoth, Seetha (2005). The Primitive Tribal Groups of Kerala: A Situational Appraisal, in *Study of Tribes and Tribals*, 3(1), New Delhi: Kamal Raj, pp. 47-55.

Singh, K.S. (1994). *People of India*, Kerala. New Delhi: East West Press.

Downloads, Chalakudy River Protection Forum (CPF) & River Research Center.
cholayar@rediffmail.com, riverprotection@gmail.com, chalakudyriver.org rrckerala@gmail.com, rrckerala@rediffmail.com.

7

Ecological Stress and Resource Conflict Confronting the Kurumbar of Attappady

S. Bindu

Introduction

The Attappady valley was not opened to outsiders till the 1950s. The Kurumbar inhabit the North West corner of Attappady adjoining the Silent Valley National Bio-Reserve with a population of 2,174 (2001 Census). There are 398 households distributed in 19 hamlets of which 13 are situated in the Reserve Forests, and 6 in the adjoining Vested Forests. The Kurumbar were perhaps the earliest inhabitants of Attappady. It is supposed that when the Badagas started colonizing the Nilgiris, the Kurumbar steadily moved to the Nilgiris' lower slopes, i.e., the Attappady valley, which is about 3,000-6,000, feet above MSL and is a high rainfall zone. They are numerically the least when compared to the other two tribal populations of Attappady. Their hamlets were originally distributed in the valley of the river Bhavani and its tributary, Varharpallam. However, all these people have their settlements inside the forests, and until recently they had no contact with the outside world.

The Kurumbar can be taken as a typical example of a shifting cultivator; non timber forest produce (NTFP) collecting tribe. They had a landlocked existence till the middle of the century. The rich forests surrounding the hamlets were endowed with numerous NTFPs, and the Kurumbar were specialist NTFP collectors for their cash requirement for market goods. The shifting cultivation practices could produce enough cereals and legumes for year round consumption (Bindu, 2008). Hunting and gathering supplemented their economy. The Kurumbar knowledge of medicinal plants is legendary and some of their reputed healers acted as herbalists of their neighbouring communities of Nilgiris and the Attappady valley.

There occurred a rapid transition in their economy since 1976 owing to the identification of the community as Primitive Tribal Group (PTG). Special Central Assistance was received for housing, drinking water, education, agriculture, cooperative society, etc. Their neighbourhood in the lower valley became a haven for migrants from Central Kerala. The shifting cultivation practices and the fertile rain fed inaccessible slopes attracted clandestine *ganja* cultivators from central Kerala and they infiltrated into almost all Kurumbar hamlets engaged in *ganja* cultivation. Their socio-cultural scenario has changed completely and their self dependency and self reliance have been depleted due to the deterioration of forests and intervention of the clandestine *ganja* cultivators in the economy. This study tries to present the ecological stress existing in the Kurumbar habitat and the resource shift in the economy of the Kurumbar.

Research Methodology

The researcher had constant contact with the Kurumbar since 1995 in connection with preparation of ethnographic notes, preparation of action plans (VIII, IX and X Plan) for development, conducting of medical camps, study of Kurumbar child rearing practices and basic personality, ethno biological study of their traditional knowledge system, etc. All the 19 Kurumbar hamlets were visited by the researcher during the last 14 years. Participant observation, interview of social functionaries, group interviews of tribesmen using questionnaires and survey of socio-economic conditions were the anthropological techniques used in the study. The Acculturative Stress Rating (ASR) scale developed for the study of the neighbouring Irular is applied in the Kurumbar case to rate the ASR.

Objectives of the Study

The major objective of the study was to present an account of the traditional Kurumbar economy. This is in order to highlight the shift that has taken place in the traditional shifting cultivation of cereals and legumes due to the introduction of *ganja* cultivation. This study also presents the ecological degradation due to teak plantation, bamboo extraction and population pressure. The study proposes to highlight the ecological degradation due to the shift in their traditional agriculture practices. To measure the resource conflicts and acculturative stress, a 20-point Stress Measurement Scale was prepared by Viswanathan and Bindu for the purpose. The study aims to present the failure of the development programmes because of the introduction of *ganja* cultivation resulting in several pathological traits.

Cultural Profile of the Kurumbar

Traditionally, the Kurumbar had a clustered hamlet life. They were living in two or three linear dwelling structures divided into compartments. Each compartment was occupied by a conjugal family. The occupants of a linear structure were usually patrilineal agnatic kins. The Kurumbar are an endogamous community, but they have had limited exogamy with Mudugar. The Kurumbar society is constituted by eight exogamous clans. The Kurumbar have an acephalous political organization and all the hamlets have their own socio-political functionaries. The headman of the hamlet is called the *ooru-moopan* and he officiates on the social, political and religious functions. He acts as the representative of the hamlet. Other important social functionaries are the *kuruthalai, bhandari* and the *mannukkaran*. The *bhandari* acts as the treasurer and *kuruthalai* performs the role of messenger and assists the *ooru-moopan* in discharging the social and religious functions.

Mannukkaran is the priest who performs all religious rites and observances for the hamlet people. He is the custodian of the *karaideivam*—the deity of the land. He is considered as a seat of wisdom of all agricultural operations in the hamlet. The socio-political functionaries wield their authority over the hamlet society. The magico-religious rites are performed by the *mannukkaran* during the clearance of wild growth, sowing, weeding and harvesting. Religious observances of the Kurumbar begin from home. The different clan deities do have specific names and are worshipped by the clan members. It is the home deity that controls their day-to-day affairs. The Kurumbar worship Hindu gods and goddesses. Lord Shiva (Malleshwara) is their principal god. Bhadrakali-hayo and Mariamma are the female deities they worship the most. The Kurumbar spend all their savings and resources of years together for feasts and festivities related to ancestral worship.

Economy

The Kurumbar had an economic life well adapted with the ecology of the area. Traditionally, they could derive their food sustenance from the shifting cultivation practices for outside goods like dress and condiments. They could earn sufficient cash from the sale of NTFPs from the rich forest of their neighbourhood. The *Kurumbar* also had the tradition of cattle rearing from which they derived the name Palu Kurumbar (Palu means milk). As they live in the forest habitat their economic pursuits involved food gathering and hunting too.

Shifting Cultivation Practices

The Kurumbar of Attappady were well known for their shifting cultivation practices or the slash and burn cultivation. The Kurumbar hamlet land is held communally but cultivation is done in separate plots (*karim kadu*) by each family depending on the work force. Cereals and legumes are the major shifting cultivation crops. Each family of a hamlet used to cultivate one-fourth of the hamlet land. Their staple food is ragi. The Kurumbar cultivate four varieties of ragi (*Elusine coracana species*) like *kola* ragi, *onda* ragi, *savarai* ragi and *valia* ragi harvesting from 3-6 months. Mixed crop cultivation is adopted. Prior to 50 years, the Kurumbar would return to the same plot only once in 12 years when the population was fewer. Agriculture calendar for shifting cultivation begins from one Mahasivaratiri (February-March) to the next Mahasivaratiri. Preparation of the plot for the shifting cultivation is the first process. The *mannukaran* is the soil expert who fixes the shifting cultivation plots for the Kurumbar families of the hamlet and allotment is usually made according to the potential of the work force available in a family. Burning of debris is the next step. Tilling of the soil with crude implements called *kothu* (hoe), *kunthali* is the next step. Dibbling sticks for sowing red gram are made. Certain seeds are broadcasted. The seeds include different varieties of finger millet called ragi, *chama* (little millet), thina which they call pandi and amaranthus species. The crop's duration ranges from three and a half to six months. Weeding is done intermittently. At the time of harvesting of crops, the whole family stays in temporary huts called *chalai* for 2-3 months in order to protect the crops from birds and wild animals. The course of action from the preparation of plots to the weeding involves a series of rituals by the *mannukaran*–the agricultural expert. The cereals and legumes collected after harvesting are dried. The drying of cereals is done in the vicinity of the threshing floor of the field hut. The dried grains are taken to the permanent huts in the hamlet and stored in huge bamboo baskets *guluma* or dug out pits in the hut called *kuzhi*. When the whole process of shifting cultivation is over, they leave the plot for weed growth for a period of three years. In earlier days, as told above, there were huge granary baskets called *guluma* that could store more than 1000 kilograms of grains. It shows that they could produce enough cereals and legumes for year round consumption. The storing baskets are made using locally available materials, such as bamboo, cane, grass, etc.

Food Gathering and Hunting

Food gathering and hunting among the Kurumbar are undertaken depending upon the availability in the forests. Numerous roots, tubers, green piths are collected from the neighbouring forests. Crabs, small animals and fish from rivers are also used as food. *Perukka, noore, cholatheeta, eratheeta, kavaltheeta, naratheeta* are the common roots and tubers available. These are collected irrespective of scarcity because of the food value and seasonal abundance. After a heavy rainfall, a variety of mushrooms and green leaves are available in the forests. They collect various types of mushrooms in different seasons, like *koyiambe, keenambae, kooleambae, puttinamabe,* and *karavand* and green leaves like *keera, mullukeera soppu, pattikkeera soppu, chowthe, vellari* are collected. *Kattarangi* is consumed as a side dish. Crabs of different species are also collected by them. Wild fruits, gooseberry, nuts and flowers are the other items collected from the forest. All these show that Kurumbar had enjoyed surplus food in the past and as such in their behaviour pattern there is no place for thrift and storing for the future.

Various types of deer, monkeys, goats, and rabbits are hunted using *kathari*–a bamboo trap. *Udumbu* (Iguana), *Veruku* (civet cat), *Kooraman* (barking deer), mongoose, *eyyu* (porcupine), *Alunku* (scaly ant eater) are the animals hunted. Birds are caught using pellet bows. *Perukku, chorapakki, kottala pakki* are the birds caught using pellet bows. They know the names of 42 species of aves.

NTFP Collection

Each Kurumbar hamlet has specific forest territories called *kadu* through which they can roam in collection of various NTFPs seasonally. With the practice of shifting cultivation, the hunting and gathering has become a part-time activity. Dammer, soapnut, medicinal plants, bee wax and honey are the NTFP collected. April, May, June are the honey collection seasons. All are transacted to the Kurumbar Girijan Service Cooperative Society. Wild roots like *padaver, nannari, chundaver,* etc., are also collected as NTFPs. Since there existed vast stretches of virgin forest rich with NTFPs, the Kurumbar Girijan Service Cooperative Society often recorded one of the highest NTFP collecting Cooperative Society of the State. The higher elevation evergreen forests were rich with honey and dammer and the forests of the lower slopes, with soapnuts and gooseberries. The Kurumbar ventures for the honey collection pursuits in a team of four or five able bodied men to the higher elevation for days together. It is considered a hard task for men. The lower value collection pursuits are family ventures and are a day time work.

Cattle Rearing and Poultry

Kurumbar are adept at rearing milk yielding animals like buffaloes, cows and goats. The Kurumbar rear cattle for milk products, not milk as they seldom drink milk. Curd, buttermilk and ghee are prepared and consumed. Certain Kurumbar families are seen nurturing drought animals of non-tribal farmers of the plains during the non-agricultural season for fattening. Goat rearing is also undertaken for sale. Some of the Kurumbar have bamboo coop for the poultry in their hamlet. Hens are raised for flesh and egg. They feed the hens with *chama*, ragi, etc. which they produce through shifting cultivation.

Ganja Cultivation replaces Traditional Crops

Along with the non-tribal migrants, illicit brewing of liquor and *ganja* cultivation also invaded the Attappady Valley. All the three tribal communities were affected. The implementation of development programmes such as housing, drinking water projects, roads, etc., brought in petty contractors to Kurumbar hamlets too. The isolated Kurumbar hamlet land in high rainfall zones has attracted unscrupulous elements among non-tribes. The *ganja* cultivation in Kurumbar hamlets became widespread after 1996. Interestingly it coincides with the introduction of prohibition of toddy tapping and liquor vending in Attappady. The clandestine *ganja* cultivators (CGC) engaged Kurumbar youths for their operations in the surrounding forests of the Kurumbar habitat. Small patches (25 to 50 cents) of forest were cleared. They selected cultivable patches of forest on the banks of brooks and streams on the slopes as the *ganja* plants require watering with hose pipes during dry spells. The CGCs operated mostly outside the traditional hamlet land. Certain Kurumbar families cultivated *ganja* plants close to brooks and streams passing through their hamlet land away from their cluster of houses. The Kurumbar were paid in cash and kind by the CGCs throughout the cultivating season. Since 1997, there occurred a visible change in the cultivation practices and the traditional crop production lessened. The traditional crop cultivation became a namesake affair for majority of the families. By 2002-03 most of the Kurumbar families started their own cultivation of *ganja* plants totally ignoring the traditional crops.

Ganja Cultivation and NTFP Collection

The staple crop has now been replaced by *ganja* and many of the Kurumbar suffer from the evil twist of the economic change. In every hamlet, few Kurumbar families act as agents of non-tribal *ganja* cultivators. They advance money prior to the cultivation season to deploy men in order to prepare the plot. For weeding, irrigating and guarding the *ganja* plants, wages are paid. The *ganja* thus raised by the Kurumbar is sold to the agents who had paid advance. Relatively easy money flows to the Kurumbar. Few Kurumbar families were able to construct concrete houses as seen in rural areas.

The NTFP collection has become less attractive to Kurumbar youths because of the availability of works related to *ganja* cultivation. Collection of the NTFPs such as honey and dammar are considered very strenuous and time-consuming. The CGS encouraged drinking habits of hard liquor which they brought to lure Kurumbar for harbouring them in their clandestine operations. There occurred a steep fall in the NTFP collection.

Ecological Stress

Population Pressure

The Kurumbar population recorded steady growth since 1981. According to 1971, 1981, 1991 and 2001 Census, it was 1,319, 1,283, 1,820, and 2,174 respectively. The population shortfall during 1971-1981 may be due to the non enumeration in a few inaccessible hamlets. Since 1981 to 2001, the population nearly doubled. The population growth from 1981 to 1991 was 41.8 per cent. A survey conducted by KIRTADS in 2006 recorded the population as 2,079 (*Baseline Survey*, KIRTADS, 2006).

The population increase indeed was a forerunner of a change in the dwelling pattern. Formerly the Kurumbar dwelling pattern was characterized by compartmentalized linear structures and clustered huts. By 2008, almost all families have independent houses and in separate plots. Most of the families prefer to have a road frontage. Similar tendencies can be seen in majority of the Kurumbar hamlets. So the sharing of the hamlet land is becoming a problem. Providing title deed to the Kurumbar is certainly going to open a hornet's nest.

Forest Degradation

The shifting cultivation practices were limited within the assigned hamlet territory. The clandestine *ganja* cultivation plots by *ganja* mafia were mostly outside the hamlet territory. They prepare the *ganja* cultivation plots by slash and burn operation. The timing of burning the slashed forests cover coincides with the burning of the hamlets' shifting cultivation plots. The fire from the *ganja* cultivating plots outside the hamlet territory create havoc in the evergreen forest during the late summer as it often generates wild fires on the slopes. This fire thwarts wild animals, and exterminates leeches from the *ganja* plots. So the *ganja* cultivators encouraged forest fires and this caused extensive damage to evergreen forests. Needless to say huge trees were burned down within a couple of years and the forest degradation took place on the Ghat section. Dammer trees of these areas also perished. This has affected dammer, honey and certain other NTFP's collection from the evergreen forests.

One can notice that the Kurumbar habitat was subjected to extensive teak plantation activities by the forest department. The teak plantation activities can be seen at Thadikundu, Anavayi, Palappada area and Gottiyarkandi, Kurukathikallu areas. The weeding operations from the teak plantation also cause destruction to the neighbouring forests. Bamboo and reed cutting from these forests were undertaken periodically for a newsprint factory. Interestingly this area is now declared as Silent Valley Buffer Zone imposing severe restrictions on Kurumbar movements.

The forests of the Kurumbar area had a marked degradation. This is obvious in major hamlets like Melethudukki, Thudukki, Anavayi, Kadukkumanna, Bhoothayar, Ghottiyarkandi and Kurukkathikkallu area which are all major hamlets. Shrubs and thickets grow even outside the hamlet land. Devoid of tall trees, the area appears to have a semi arid look in summer. The hamlet people complained about summer heat now felt in the habitat. Elders complained about the climate change and the water shortage. This aspect is mostly reported in Ghottiyarkandy, Kurukkathikallu and Bhoothayar area which is a contiguous stretch of dry areas of Attappady.

Drinking Water Shortage

Over the years from 1971 to 2001, the Kurumbar population has increased. The Forest Department's teak plantation activities could alter the quality of forests. There occurred disputes for land between Forest Department and the Kurumbar in Palappada and Anavayi hamlets for encroaching land. There were similar instances in Gottiyarkandi area too. *Ganja* cultivation in forest patches outside hamlet territories destroyed the forest. The degradation of forests on the banks of brooks and streams reduced the water flow. Now almost all Kurumbar hamlets suffer from shortage of drinking water as these once perennial source of drinking water got dried up. Immediately after rains, there occurred a recurrence of diseases due to water pollution. Menace of rats, destruction by wild boars, marauding elephants, all spread havoc on Kurumbar cultivation and destroy their stores of food items, etc.

Acculturative Stress Measurement

The Kurumbar contact with non-tribals, indeed were of many kinds. The Kurumbar make overnight halts in the growth centres whenever they visit for the marketing needs. Thus, they get exposed to film viewing, drinking and smoking. Plantation workers, *ganja* cultivators, petty contractors and labourers associated with development work make night halts with the Kurumbar families. Visiting forest officers, anganwadi teachers, and teachers of L.P. Schools constitute another category. A number of Kurumbar make distant travels to enrol their children in distant MRS and Ashram schools.

The Kurumbar society had been subjected to tremendous socio-political changes. The Panchayati Raj System and law enforcing machinery of the mainstream society have replaced their traditional social functionaries and social control mechanisms. In their transactions with non-tribes an exploitative situation revives. Transaction of agricultural produces, wage labour payments and various exploitations occur. These situations indeed cause stress and strain among the Kurumbar.

The 20-point scale prepared for the study of Irular had been applied in assessing the acculturative stress of the Kurumbar. The significant fact is that the Kurumbar themselves have rated the factors causing the greatest stress in their lives, the factors that do not cause them any anxiety at all, and those factors that fall in between these two ratings. An analysis of their ratings shows that they have the highest level of stress in relation to (i) eco-degradation, (ii) loss of traditional livelihood, (iii) intrusion of pathological traits, alcoholism, and boot legging into their midst, (iv) loss of traditional social control mechanisms and control over political system, and (v) circumstances that lead to loss of confidence in government machinery and the State (self-rated score is six in all these cases). Significantly, they have reported absolutely no stress (self-rated score is zero) arising from conflicts with non-tribes and settlers or from any sense of having low social status or marginalisation.

Similarly, contrary to shifting cultivation, brewing liquor or tapping toddy were not traditional practices of Kurumbar. *Ganja* cultivation accelerated the clandestine *arrack* distillation. The Kurumbar had fallen prey to these vices. *Ganja* and liquor addiction resulted in neglect of family life, affliction of new diseases, *ganja* psychoses, liver *cirrhosis*, suicide etc.

Their loss of confidence in the State and administration can be traced to the number of court cases with forests and excise officials and the failure of the development schemes of the Kurumbar that they attribute to official apathy. They consider that the Government officials were not able to understand the traditional ways of land assignment, landholding pattern, and customary rights, and this lack of understanding has seriously hampered them. The distress in this connection is also extremely severe. In the case of the next factor causing high stress, viz., loss of traditional social control mechanism and control over political system, it is to be noted that *Oorumoopan* was the sole authority in the past and nobody dared to make drunken brawls or harbour non-tribes without the permission of the *Oorumoopan* because of the fear of punishment that would be inflicted upon them by the social functionaries. Non-tribal *ganja* gangs are today seen staying with certain families without the permission of *oorumoopan* or other social functionaries. The social control mechanism has become totally ineffective. The situation is rated as extremely severe and hence the score is 6.

Similarly, the economic transactions with the non-tribes are always exploitative in nature. Therefore, there is very high stress here too. The rated score is 6 in this also.

Unlike in the case of many other tribes, the Kurumbar do not have stress related to land alienation as their hamlets do not have settler occupation of land, mainly because the land is inside the

Reserve Forests. The zero score should be understood in this context.

Stress arising from hostile situations with non-tribes and settlers is also absent (zero score) due to their habitats being located in the Reserve and Vested forest region. However, over the years, with increased interaction with the settlers, the Kurumbar society's distinct identity has been pushed to the lower strata of the society. But, as they are not dependant on the non-tribes for employment, and as employment is available in the hamlet itself, they do not find such issues of low social hierarchy stress-inducing (ASR rating is 0).

Nevertheless, their marginalized role in Panchayat Raj Institutions has caused them high stress (Score: 5). The Kurumbar hamlets located in isolated pockets are hardly aware of the *oorukoottam* (hamlet assembly) constituted by the Grama Panchayats. There is minimum participation by Kurumbar and often they take no interest. There exists a wide gap in the traditional tribal councils and the present *oorukoottams*. It has been evaluated by them that the *oorukoottams* benefit the political functionaries and not the hamlet people. Equally stressful has been the fact of their inability to keep pace with the technological advancement of the mainstream society that leads to an increase in the gap between the two. Loss of traditional livelihood, language, labour opportunities, and social life, and economic pressures create some stress, but do not perturb them excessively.

Conclusion

For the past one decade there has been transition among the Kurumbar from their traditional agricultural practices to *ganja* cultivation to a great extent. They reason that a lot of energy input is involved in shifting cultivation, whereas in *ganja* cultivation, there is a minimum amount of energy because of the higher price and windfall cash. The isolated Kurumbar hamlets and high rainfall are suitable for *ganja* cultivation. The after-effect is that the traditional shifting cultivation practices are affected. There is erosion of traditional knowledge involved in shifting cultivation because the non-tribes are mainly involved in *ganja* cultivation and the Kurumbar have become their labourers. This is also an indirect method of resource alienation which the Kurumbar are not aware of. They are also unaware of the money made by the *ganja* mafias. Alcoholism and associated diseases, suicides, and crimes have become rampant. Deforestation occurred outside the hamlet land due to clandestine *ganja* cultivation, 90 per cent of the money in *ganja* cultivation taken away by the non-tribals.

The Kurumbar economy had a pathological deviation during the 1990s. The *ganja* cultivation had caused extensive damage to forest of the area. The Kurumbar themselves started cultivation of *ganja* plants in the hamlet land. Many a Kurumbar got addicted to *ganja* smoking and spurious liquor. In the mainstream economy, the Kurumbar earned the tag of '*ganja* cultivators' as they found a different use for their land. The extensive damage to forest of the neighbourhood resulted in the drying up of drinking water sources, loss of fertility of land, scarcity of NTFPs in the vicinity, etc. Lately, Government has taken to stop the *ganja* cultivation and majority of *ganja* gangs have withdrawn from the scene. The Kurumbar have lost their means of subsistence. Nowadays they derive their sustenance from house construction works.

Earlier the Kurumbar had a clustered hamlet life. Now the hamlet houses are dispersed. The educated among the Kurumbar are interested only in the development of homestead plots focusing on cash crops. Currently, as we find in 2009, the Kurumbar baffled with the loss of balance of their ecology, and the vast reduction in the income from the *ganja* cultivation. At the same time, their food production and NTFP collection are inadequate for year-round sustenance. However, there is a ray of hope as literacy among them has been steadily increasing and more youth are showing interest in higher education. This human resource, coupled with the natural resources of the Kurumbar area, can pull out the Kurumbar from the abyss into which they had fallen.

REFERENCES

Bindu, S. (2008), *Action Anthropology: Tribal Development* Issues, Indian Indigenous Peoples Service Society, Wayanad, Kerala.

Kirtads (2006), Baseline Survey of the PTGs of Kerala, Unpublished report, Kerala Institute for Research, Training and Development Studies of SCs and STs, Kozhikode, Kerala.

8

Changing Ecology and Changing Tribal Communities of Wayanad

A Case Study of Displaced Tribals on Account of Karapuzha Irrigation Project of Wayanad District

P.V. Mini

Introduction

Human settlements of the tribal areas have been gifted with rich mineral resources and bio-diversity. With the development of science and technology these natural resources are being increasingly exploited in the name of the prosperity of people and country as well. Numerous projects have been launched for the management of water, mineral and forest resources in the tribal areas both for the development of the tribal communities and the areas as such. These projects are mostly located in tribal areas which are endowed with rich raw material and natural resources. The local inhabitants are displaced due to acquisition of agricultural lands or house sites, and consequently their economic base and socio-cultural life are disrupted. This process of taking away the resources enjoyed by them started during the British period which affected their life stage by stage, and can be viewed as blatant exploitation of their life and resources.

The execution of reservoir projects in India through Five Year plans has received a great importance in the post-independence era. In a period of three decades about 600 major or medium dams have been constructed in the country. With the establishment of major projects, the local tribes are largely affected and sometimes uprooted from their traditional moorings causing socio-economic problems. Cash compensation for lands acquired in tribal setting has a devastating effect, as it hardly does any good to the tribal who is not good at money management. Further, it results in loss of economic resource besides disturbing their age old customary rights and privileges.

In Kerala, some tribal communities have been displaced due to various reasons like the construction of dams for irrigation projects or for power generation and establishment of wild life sanctuaries in their traditional hamlets. As a result, the flora and fauna chain of the virgin areas undergo serious disturbances, causing ecological imbalances. The tribal communities, who had been depending on their minimum need economic operation of hunting and gathering for food, shifting cultivation, sometimes settled cultivation, etc., also had to suffer from these ecological changes.

In this chapter an attempt is made to show how a medium irrigation project called Karapuzha Irrigation Project constructed in a tributary of a major river in Wayanad district in Kerala has affected the life and culture of a group of tribal communities who had been living there without disturbing their ecosystem. This is a case study of a settled agriculturist community and how they have to discontinue their agricultural practices, change their belief systems and make livelihood adaptations.

Methodology

The data for the present study has been collected from the project area, where the tribes like Mullukurumans, Paniyans, Kattunaickans and Vettakurumans traditionally inhabited. The case study has been prepared from the Mullukuruma tribal settlements of Nellarachal and Attuwayal of Ambalavayal panchayat of Sulthan Batheri taluk of Wayanad district. Details of the project plan have been collected from the authorities of the irrigation Department in Wayanad. The organisers of the Karapuzha Adivasi Avakasa Samrakshana Samithi, who have been protesting to secure their legitimate right of rehabilitation measures, have also been interviewed. Field visit to take photographs and to evaluate the present situation have also been conducted. An interview guide was prepared and administered to a focused group to elicit information.

Background of the Project

The Wayanad district is drained by Kabani and its three major tributaries, viz., Panamaram, Manathawadi and Thirunelli. Bavalipuzha and Noolpuzha are the other tributaries. Kabani river is unique because it is one of the three east flowing rivers in Kerala and it is an important tributary of the river Kaveri. The Panamaram tributary river originates from Lakkidi and has five tributaries, viz., Kavadampuzha, Kadamanthodu, Venniyode Puzha, Karapuzha and Narassi Puzha.

Wayanad has a population of about 7.86 lakh of which 90 per cent depend upon agriculture for sustenance. There are 40,129 farmers, 74,813 agricultural labourers and 17,413 plantation labourers in the district. Another 37,267 people earn their livelihood from animal husbandry and forest produce. Wayanad as its name indicates, i.e., 'Vayal Nadu', which literally means land of paddy is famous for paddy cultivation. The farmers mainly depend upon rainwater and other sources for paddy cultivation and as such they have to limit their crop for one time cultivation. In order to raise the paddy into double crops, irrigation facilities have to be provided to the farmers. In 1980, there were 30,000 hectares of paddy fields in Wayanad. It has shrunk by more than 76 per cent to 7,000 hectares at present. The Karapuzha irrigation project, envisaged to irrigate 5,221 hectares of land for raising two crops of paddy and other crops like banana, ginger, vegetable, etc., has been launched in the year 1978 with an estimate of 7.6 crore. It has been planned to construct an earth dam at Vazhavatta across the Karapuzha river and to have two canals for irrigation. The location of the storage is at Vazhavatta which is 6 km away from the Kalpetta-Sulthan Bathery Road at Kakkawayal. Currently this project is redesigned as a multi-project to serve the purposes of drinking water, small hydel projects, tourism development, fisheries, etc. After 30 years the project is still incomplete and has not realised any of its objectives, for the water flowing through left bank canal now leads only to the Kabani River. However the earth dam has storage of water after its partial completion in 2005.

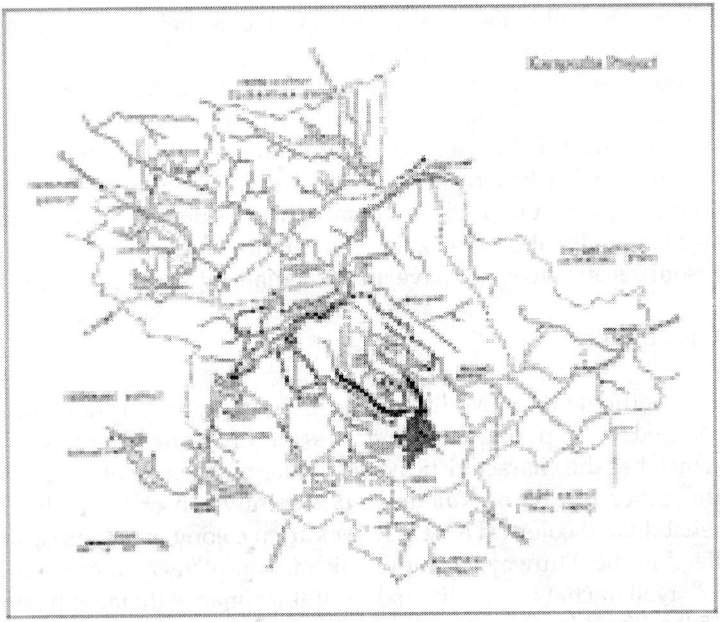

Fig. 8.1: Karapuzha Project

Nineteen tribal hamlets have been affected by the Karapuzha irrigation project. Some of the traditional hamlets have now been fully submerged under water and some of them are partially submerging. It is observed that when the water level rises, it submerges these and when recedes it is exposed and some of the families still continue to live in the project area. According to the Government sources, since the year 1978, 218 families have already been evicted from the project area. Out of them, 84 families have already been rehabilitated by providing 75 cents of land and houses. The remaining families are yet to be rehabilitated although very recently the Government has promised the protesting organisers that they will be rehabilitated with the same extent of land as the 84 families and an additional amount Rs. 3,000 per family for erecting temporary houses.

The livelihood pattern of the tribes living there are entirely different. Mullukurumans are settled cultivators; Paniyans are agricultural labourers, Kattunaickans recognized as a primitive tribe are hunters and gatherers, and Vettakurumans are artisan tribes. Here, all of them have been affected as they were uprooted from their traditional ecosystem. The Mullukurumans were dispossessed of their land for paddy cultivation, the Paniyans lost their agricultural labour, and the hunting-gathering of Kattunaikan were reduced due to depletion of resources and the artisan tribe Vettakuruman lost raw materials. Except the Mullukuruman community, no other tribal communities resided in the area had wet land cultivation. Thus, this community had lost its livelihood resources and does not have an immediate alternate resource.

Case Studies

Two settlements of Mullukuruman community that have been selected for preparing the case studies are—Nellarachal settlement, which has the characteristics of a submerging colony at times and the other one is of *Koyilonkunnu* colony which is a typical rehabilitated colony. These Koyilonkunnu colony members earlier lived in the Attuwayal colony which is completely submerged.

Nellarachal is a traditional Mullukuruman settlement having 55 families. Name of the colony originated from the word '*Nellara*' which means a granary of paddy and it is believed that the Mullukurumans of this area took land from the Nilambur Kovilakam for paddy cultivation and the place, where large amounts of paddy of the Kovilakam was kept, came to be known as '*Nellara*'. Mullukurumans of this settlement were expert paddy cultivators who used varieties of rice. They know both wet and dry land cultivation. In dry land, they cultivate coffee, banana, pepper, turmeric, ginger, vegetables, etc. They were also proficient in 'slash-and-burn' cultivation. They raise crops like ragi, chama, root vegetables, chillies, etc. To support their cultivation they kept large number of cattle for ploughing and grazing. There was enough fodder for the cattle and the cow dung could be used for cultivation. Milk and milk products also provided their own consumption and to outsiders as well which generated an additional income.

Their traditional houses were thatched with the locally grown grass available in their paddy field and they could easily get the materials for thatching houses.

All the members, both male and female, had enough work on their own land and there was no famine. According to their custom, they celebrated the *Thulam Pattu* of the Malayam era *puthari uthsavam,* a harvest festival. On this occasion, they cooked the newly harvested paddy cultivated by them, and this is considered as a big event for them. Besides, to celebrate *Uchal* which falls on *Makaram* 28th of the Malayalam era, they went out for hunting, preferably a rabbit, which is then ceremoniously cooked and consumed by one and all. Hunting is a competitive event in their life and it can be observed that a bow and three arrows are kept besides the dead body of their males. "Thus a Mullukuruman is born, lives and dies as a hunter" (Misra, 1971) .

All the above characteristics of the Mullukuruma community have been completely disturbed by the irrigation dam project and the consequent loss of land and the ecosystem, which provided everything for them. Their paddy land was submerged and consequently they lost their occupation, and now they have to get their rice from a ration shop. They have lost their cattle and manure. There is no wet land and slash-and-burn cultivation and no paddy for *puthariuthsavam* and no grass for thatching their houses. They could not get fuel for cooking purposes. The hunting practices had to be stopped mainly because of forest loss. Even the settled cultivators became agricultural labourers. Only the younger generations are going to the other parts of Kerala, and to the Karnataka State in search of employment. Their womenfolk who have been mainly assisting the paddy cultivation at all levels have become unemployed. Culturally the womenfolk do not go out to do agricultural labour or for any other employment. The older people, who did not diversify their skills, have become the worst victims of unemployment. The rise of the water is always a threat to their habitat, and as such, they have been thrown to an uncertain life. All these changes have reduced their income that has affected their development in all respects.

Koyilonkunnu Colony

It is a typical rehabilitated colony. The original plan was to accommodate four families in a house. As the beneficiaries protested, it is now changed to two families in one house. There is only a small kitchen and they have a common chimney.

Currently there are 44 families living in 22 houses. But the houses constructed for them are of low quality. The RC houses are found to be leaky. The windows and doors are shaky. A few houses are found vacant and some of the houses are used as cattlesheds. Recipients of these houses have gone back to their original houses, which in fact, are in a submergible area. In this colony, *daivapura,* which is typical of every Mullukuruman settlement, is also constructed. There is water scarcity and they have to go down the hill to fetch water. There are no shops to purchase goods for their daily living. Land is infertile for cultivation. There is no scope for fuelwood collection.

Fieldwork has revealed that only those who have no other option are occupying the rehabilitated houses. On enquiry, it has been found that the land for this colony has been selected from the acquired portions, where all the trees have been cut, and as such the land has already become barren. Therefore, the main reason for leaving their rehabilitated houses is that they have no chance to meet their livelihood, whereas, their original ecosystem provided all the things for their livelihood. The rehabilitated area has become an alien area for them. Their attempts to find alternative means of living, like ginger cultivation, have also failed because of the non-conducive weather conditions. In fact, they have to starve during the rainy season.

Conclusion

There have been both large and small irrigation projects implemented for the benefit of farmers in Kerala. The induced development, though brings progress in general, may affect the displaced people negatively. But at the micro level, benefits of such schemes did not reach the marginal farmers. An unimaginative approach of the Karapuzha project has failed to achieve its main objectives. It has taken away the sole livelihood resource of paddy cultivation from the community living near the downstream. Indigenous knowledge system of the Scheduled Tribe Mullukuruman community is found to be of no use to them now. The bio-linkages of cultivation have also been broken.

Ecology shapes the human livelihood patterns and it is reflected in their social and cultural aspects also. The tribal communities, who have been found to coexist with the ecosystem, would be worst affected if ecology is disturbed in any way. Even in a small project, as in the Karapuzha irrigation project and its rehabilitations, the tribal people have faced untold sufferings and threats to their very survival. An understanding of the communities living in a particular ecosystem and how much they can adapt in a new environment has to be well worked out to avoid such violation of human rights and to enable these populations to assert their right to live with dignity.

REFERENCE

Misra, Rajalakshmi, (1971). *Mullukurumbas of Kappala,* Anthropological Survey of India, Calcutta.

9
Tribal Development, Land Alienation, Displacement
A Case Study of Karapuzha Dam Project

A.M. Prasad

This chapter is of land alienation, displacement, and compensation for loss on account of the Karapuzha Dam Project in the Wayanad district of Kerala and a critique of tribal development. Karapuzha project area is my homeland and this is an insider's view of the experiences of the tribes.

Karapuzha project is the first irrigation project taken up in Wayanad district. Wayanad with an average elevation of 780 metres above mean sea level, lies on the eastern slope of the Western Ghat and is inhabited by several hill tribes, and settlers from different parts of the State. But most of the populations are rural, and are engaged in agricultural activities. With a view to bring more and more area under paddy cultivation by irrigation to overcome the deficiency of foodgrain production, the possibilities of implementing a few major irrigation projects had been explored in the Wayanad area, out of which the Karapuzha project was found to be the most feasible one and was taken up for execution during the First Five Year Plan. The project is to construct an earth dam at Vazhavatta in Vythiri taluk. This is purely an irrigation project which plans to construct a storage reservoir and canal system to irrigate a net area of 5,221 hectares for two crops. Construction was started in February 1977 during the Fifth Five Year Plan.

The revised project report by incorporating all the updated design changes and by adopting the 1999 schedule of the rates has been prepared for an amount of Rs. 253 crore. The work is not yet completed. An independent division called 'Karapuzha Project Division' started functioning from February 1977 onwards with headquarters at Kalpetta. Actual commissioning of the project was to have taken place in 1984-85. This did not materialise due to meagre placement of funds from 1980-1981 onwards. The target date for the commissioning of the project was fixed at March, 2004; provided adequate funds were made available. In addition to the originally designed scope of the project, a new proposal for 0.5 to 0.75 MW of power generation and drinking water supply for adjoining towns, and for tourism, are under consideration of the concerned departments, for which primary investigation has already been conducted.

The work of the project started in the year 1977. The project area covers Meenangadi and Ambalavayal Panchayaths of Sulthan Bathery taluk and Muttil and Muppainad panchayats of Vythiri taluk in Wayanad district. It covers an area of 5,221 hectares. The victims of this project are mainly some tribal groups, namely Mullukuruma, Paniya and Kattunayakan. They had been staying in the area, which is now submerged. The first resettlement package was only for one settlement. The project-affected persons of this settlement were resettled in Nedumbala at Muppainad Panchayat. They received the compensation of Rs. 5,000 per acre. The next resettlement package was introduced in the year 2002 onwards. But this time the compensation package was at the rate of Rs. 6,000 per acre. As a part of this second package, a large number of settlements in the project area were resettled on two hill tops. One small building has been provided for four families. One hill top is provided for the Mullukuruma groups and another one for Paniya groups. People of eleven Mullukuruma colony were rehabilitated, viz., Vellada, Atuvai, Nellara, Nhamalam, Palavayal, Naduveedu, Thammiyadu, Erangheri, Kuzivayal, Narakakolly and Chanakolly. Four Paniya colonies, viz., Cheeprum, Naikolly, Nellara and Madumkunnu were also resetlled. A few non-tribal families also lived in this area. These families were resettled later by providing 4 cents of land in the nearby area. A point to be noted is that the non-tribes got compensation for their lost land at the rate of Rs. 1 lakh per acre.

There are only six or seven Mullukuruma families and some

Paniya families living in the resettlement area now. I enquired about the other displaced tribal families who are no longer there and found out that many had left as the housing condition in the resettled colony was poor. The houses look as though they might fall any moment. The windows are covered with cardboard pieces. One housing unit is made up of a small bedroom, kitchen, and verandah. These rooms are so small that only one person can move about in a room at a time. During the monsoon, it is just impossible to live in these houses.

All the case studies that I did in this area lead to this conclusion. For example, take the case of Pradeep, 20 years old Mullukuruma in Nellerachal colony has had absolutely no benefit from the project. He is technically displaced but he refuses to leave his old home in the project area. In the new settlement allotted to him, there is no paddy cultivation and no cattle. And the house there is just about to fall down. Balakrishnan (58), Mullukuruma, is also in the same plight. He is also bothered by the fact that he cannot do any Kuruma ritual in the new settlement. All his ancestors and gods are in the submerged area. Parvathi, 45 years old, Paniya, also refuses to resettle. There is no proper housing, work, or drinking water in the new place. Velli (46), Paniya is an expert in paddy cultivation, but is not skilled in any other kind of work. He feels lost in the new settlement and has come back to his old home. Balan (57), Mullukuruma, Achuthan (68), Mullukuruma, Ramadas (64), Mullukuruma, Leela (40), Paniya, Bharhathi (35), Paniya, Vijayan (40), Paniya, all have the same story. Their lives are threatened, their future is uncertain, and their communities fragmented.

Conclusion

Due to the fragmentation, they have lost a sense of belonging. They have lost the traditional rituals they treasured. They feel the loss of unity and their own identity, as a whole. The amount of fertile land diminishes the opportunities for cultivation, and the paddy fields are no longer in their new place, and, they have to buy rice. Their daily wage system does not allow them to have rice every now and then; it affects their health. The development of a tribal community is not just possible by giving them some land, but only by treating them as equal citizens. Development certainly needs sacrifice, and the common good is important. Individuals and communities have to co-operate for the sake of development, but the question is: why is it that some communities are always the chosen ones to make the sacrifice?

10
Impact of Affirmative Action and Cash Crop Cultivation on Koraga in Kasaragod, Kerala

S.K. Shashikumar

Reservations–as they are called in common parlance in India–are an integral part of the Constitution of India. On the one hand, it acts as the compensatory discrimination for the group of people who suffered from exclusionary and discriminatory practices, while on the other, it is said to be creating the 'other' from within.

Impact of affirmative action or reservations on historically marginalised communities is a major issue being debated in India these days. With a glance on the Koraga inhabiting two Panchayats Kumbdaje and Badiadka of Kasaragod district, let me touch upon these debates.

In this chapter, I will look into the impact of States' protective discrimination practices (for the marginalised communities on one hand) and the possible reasons that lead to very little success of such programmes. I myself will remain confined to analysing the impact of reservations in political spheres particularly in local politics/political parties while analysing the impact of reservations and I will concentrate particularity on rubber plantations while looking into the impacts on Koraga's traditional occupation.

Koraga is a Scheduled Tribe living in an area called Tulunaad, which is constituted by the areas now coming under two south Indian States–Kerala and Karnataka. Now, for administrative purposes, they are called the Primitive Tribal Group in both the States. They claim succession of King Hubashika, but majority of them are depending on basket-making for survival.

With 422 families comprising 1,572 members of which 800

males and 772 females, as per the staff of the State Health Services Department, they are one of the major social groups with highest incidence of Tuberculosis and addiction to chewing habits.

March 2008 is memorable for the inhabitants of ST (Koraga) colony near Badiadka because one of the inhabitants of this colony, this time, passed 10th standard examination. Major local dailies had given importance to this, the local population appreciated her. Only after 60 years of Indian independence, a Koraga girl in this settlement could pass the 10th standard examination.

Kasaragod district is identified as one of the most backward districts of Kerala. Seven languages are spoken in the area. The areas north of the river Payaswini vary from rest of the areas of the district, in terms of language, infrastructure, and geography. Cashew and areca were the major crops in the area. Koraga is one of the Scheduled Tribes in the area. Jainism was also a major religion here like Hinduism, Islam and Christianity.

There were not too many churches in and around Kasaragod in which languages other than Konkani were used for Sunday prayers, but now there are a good number of churches using Malayalam for those, who have come from the southern regions of Kerala.

The shift to cash crops is not a new matter in the area. A couple of decades ago there was a major shift, when arecanut got a high market price. Almost all paddy fields were converted into areca fields, but this in a sense did not make much direct impact on Koraga. However, some *dalit* populations were the prime sufferers.

The shift towards cash crop cultivation, particularly rubber, is a new trend in the area. Lands that remained uncultivated were the rich sources of raw materials for basket making. With the arrival of new populations from southern parts of the State, uncultivated barren lands started getting market value and local people started to sell out their lands.

Earlier rubber plantations were owned only by persons belonging to Christian communities, who arrived from southern regions of the State, but now the situation has changed; the native population with as much as fifty cents of land take tedious efforts to plant as much rubber as possible in the available space. Introduction of new cash crops has brought tremendous change in the land-usage pattern in the area. Available pieces of land are now turned into rubber plantations. This has shrunken down the area that used to provide the raw material for basket making. This, in a way, has affected the illiterate Koraga, who like to continue with the traditional occupation of basket making.

Apart from discussing the negative effects of the cash crop cultivation on Koraga population, I believe that I need to see whether the new trend has accommodated the marginalised populations like Koraga or whether it has given any room for the participation of populations like Koraga.

An attempt to trace the ownership of these plantations reveal that none of these plantations are owned by Koraga. I believe that aversion towards new symbolic capital embedded with production in capitalist mode is not the reason for this, or to put it in other terms, economic backwardness is the barrier for this. It can be said that the sixty years of protective discrimination has hardly helped the community to accumulate some amount of economic capital.

Impact of rubber plantation on the traditional occupation of Koraga is in two ways. On the one hand, there is the disappearance of sources of raw materials required for the traditional occupation, and on the other, the traditional baskets are now being replaced by baskets made of rubber and other synthetic materials.

In this context questions like why the community, in spite of all these preferential treatments, suffers from the cash crop cultivation of neighbouring communities, why it fails to represent anxieties before the State, or do preferential treatments really empower the marginalised communities like Koraga, do such policies really protect the rights of the populations in threat situations, etc., become significant.

Failure of policies for the disadvantaged communities was earlier attributed for the negative response from them. As Hall and Midgley (2004) put it, the underclass phenomenon is no longer attributed to low intelligence or culturally transmitted negative behavioural traits, but to wider social causes. Poor people are not held responsible for their deviant actions, but are viewed as the victims of extrinsic forces.

Protective discriminations have helped some of the members of ex-untouchable groups to compete for seats reserved for them. Among the Koragas, there are persons working in government services. But how far did such policies help them to mobilise the resources around them? There is not even a single Koraga among the leaders of major political parties in Kasaragod today.

Protective discrimination has helped historically

marginalised communities to some extent to represent themselves before the State. But it seems that among the Koraga, it did not go beyond that, even after fifty years of independence.

To conclude, reservations in decision making bodies are an integral part of Constitution of India. Koraga being one of the Scheduled Tribes can be said to be benefited from this, up to an extent–representing for the sake of representing. However, it (the participation in decision making) did not mean anything.

Both Universalist citizenship and particularist rights are parts of entitlements offered by the nation. With Universalist rights, the communities that have already acquired some amount of social and cultural capital manage to advance themselves easily but those who receive the benefits in the name of particularist rights, but have no other capital have to initially accumulate some amount of capital. In a way in the name of Universalist rights, the State seems to support those who have already accumulated some amount of capital. This seems to be true as far as the Koraga in Kasaragod is concerned.

REFERENCE

Hall, Anthony and James Midgley (2004). *Social Policy for Development*, Sage Publications, New Delhi.

http://kerala.gov.in/economic_2007/chapter14.pdf

http://kerala.gov.in/keralacallaug_07pg28-29.pdf

http://planning.kar.nic.in/khdr2005/English/Main%20Report/10-chapter.pdf

http://www.indiatogether.org/2005/apr/socrsvstudy.htm

http://www.wsws.org/articles/2006/jun2006/indi-j03.shtml

11

Ecological Adaptation and Inland Fishing of the *Padanna* Community of Kerala

V.S. Subash

The Kurumaly and Mannaly rivers originate from the Western Ghat in Thrissur district and joining at Palakkadavu near Arattupuzha, it becomes the Karuvannur River. This again is joined by the river known as Canoly Canal from Trippayar to Chettuva estuary. This entire river is around 65 Km. in length. The river influences the lives of the people who are living in and around its basin, through watering of the paddy fields and the lands and maintaining the fertility. Besides, it is a source of other resources, such as varieties of fish, sand, mud, etc. Earlier the river was famous for inland transportation which even extended up to Kollam district. When we travel through the Karuvannur river and related wetlands (lagoons), we can see the entire habitat of *Padanna* community on the edges of the river and in the related vast paddy fields. This is one community which developed a way of life and economic activities centering on its environment. In this chapter an attempt is made to portray how the inland fishing of the *Padanna* community of Kerala was an ecological adaptation and how the changes brought to the ecology by human interventions have destroyed the livelihood of the *Padanna* community.

Ethnography of *Padanna*

The *Padanna* community is concentrated in the Taluks of Thrissur, Mukundapuram, Kodungalloor and Chavakkad of the Thrissur district, and also in the adjoining Ponnani taluk of the

Malppuram district. The river Karuvanur borders the above taluk in Thrissur district. Geographically the district of Thrissur can be demarcated into three distinct regions: hill areas, plains and coastal area. The plains cover about fifty per cent of the total area of the district. About sixty per cent of the plains are one to three metres below sea level. These areas are utilized for paddy cultivation from October to May, while the fishing continues throughout the rest of the year.

Etymologically the word *Padanna* means a 'Small Island' in the river, and in its adjoining areas. The salt panning area is also known as *Padanna*. The community called themselves as *Padanna* or *Padanna Kannakkan*, but the local people call them as '*Kannakkan*' although the *Padanna* community has no relation with the *Kannakkan* community, the latter are concentrated mainly in Kozhikode, Malappuram and Ernakulam districts, and in some parts of Palakkad and Trissur districts. Their primary traditional activity is plucking of coconut. In Trissur District, the Kannakkan population is numbered as 34,039 by 2001 Census. But actually the total population of Kannakkan community in the district is estimated to be 2000. The higher number in the Census report has resulted from the mistaken entry of *Padanna* as *Kannakkan*. According to the 2001 Census the *Padanna* community is enumerated to be only 4,519, far less than the actual figure, which is approximately 29,000.

The Ecological Adaptation of the *Padanna* Community

The *Padanna* Community's way of life and the occupational patterns are very much adapted to the particular ecological system. The main traditional economic activities of the *Padanna* community were fishing and inland transportation. They also engaged in certain paddy related works such as pumping the water in the paddy field by using sixteen to twenty-two leaf wooden wheels as a secondary economic activity along with the Pulaya Community. Men folk mainly engaged in fishing, inland transportation, and water pumping, whereas women were engaged in selling fish, coir making, and agriculture work such as planting the paddy saplings, plucking the weed, and harvesting.

Fishing

Organisms and environment form an interactive system. Since the *Padanna* people are living in the edges of the river and the *kol padam* (wet lands), fishing forms an important occupation of the community. In this particular ecosystem, the community themselves is adapted to their own technology for utilizing these natural resources. Therefore, they have a rich knowledge about the climate and the related conditions of the river, availability and variety of fishes in different seasons and places. They have four types of methods to catch fish, which are as follows:

1. *Veesal Vala* : A round shaped net with lead on the edges, and is used to catch the fish from a *Vanchi* (country boat) and from the shore. Different sizes of nets are used to catch different types of fish.
2. *Kandadi* : It is a kind of a lengthy net which extends between 20 m to 100 m. It is usually tied across the river, canal and kol padam.
3. *Pathazham* : It is a kind of trap made of small thin reapers made of arecanut tree. It is installed across small canals or river-sides to catch small fish.
4. *Kolli Choonda* : A type of hook used for catching large fish. More than fifty hooks are tied in a long string, which is placed across the river.

The *Padanna* people are traditionally masters in making all these kinds of tools. Two types of *Vanchi* or canoes are used for catching the fish, viz, *Kettuvanchi*, and *Tharavanchi*. *Kettuvanchi* is made up of a cheap wood like *Punna, Kasumavu, Mavu, Aanjili* and other locally available cheaper woods. The woods are tied together *(Kettuka)* in a particular manner by using coir. *Mullamkayini, Palkayini, Mayilellu, Pongu*, and *Maruthu*, etc., are used to make *Tharavanchi.* Instead of coir, brass rivet is used to join the wooden plates. The life of *Kettuvanchi* and *Tharavanchi* are more than ten and eighty years respectively.

Inland Transportation

Geographically, people in the area are isolated by the River and *kol padam* which are paddy fields closer to the sea where salt water intrusion occurs and people alternate between rice and fish cultivation. The transport and other communication facilities depended on *vallom* and *vanchi*. The former was used for transporting goods, whereas, the latter was used for people's transport. A *vallom* has capacity to carry more than 25 tonnes. The rowers of these *valloms* were exclusively from the Padanna community. They are considered traditional experts in the field although most of the *valloms* were owned by people from other communities. However, the *vanchi*, the small boat, which is mainly used for fishing, is owned by individuals of the community. Even though the big *valloms* were owned by other community members, the rowers were not being considered as labourers of the owner, but the *valloms* were taken on rent or lease by the *Padanna* rowers. Similiarly, fishing is also considered as an independent work.

The community members are also engaged in paddy cultivation as agricultural labourers. This is only a secondary economic activity, due to the water logging situation of the area, which is a peculiarity of the ecosystem. Fishing and inland transportation were the major sources of livelihood of the *Padanna* community.

Changing Life of the *Padanna* Community

In the earlier period cotton twine was used to make the fishing net which was purely a product of manual labour. The life span of net was limited too. Later on, nylon twine and machine-made nets were introduced, and these nets became available in market. At present, Persian nets are also available in the market. No other mechanization processes are introduced in thc fishing till now. Introduction of the nylon and Persian nets helped the community to better exploit the fish wealth.

However in the earlier days, fish was abundant throughout the year, even if the river became dry between the months of April to May. Also, the fertility of the paddy field was very high as the river deposited silt through the entire field during the monsoon. It also prepared the ecosystem for the reproduction of the fish varieties. Later, the government implemented irrigation facilities through constructing bunds, check dams, dams and reservoir, etc., especially the construction of Chimminny Dam at the source of the Mannaly River. This blocked the natural flow of water into the paddy fields during the monsoon season and the seasonal migration of fishes to the paddy fields for their reproduction. Mechanization of paddy cultivation and the introduction of chemical fertilizers and insecticides adversely affected the entire ecosystem. As a result, the fish wealth reduced and some demersal species (fish generally inhabiting the bottom layers) are extinct. Species such as *Aral, Pattathi, Kadu, Mushu, Kallaral, Prayi, Puzhan,* etc., met with such a fate. And other pelagic species (fish generally inhabiting the surface waters) like *Mundathi, Bral, Karippidi,* etc., are afflicted with diseases.

Construction of bridges has diminished the inland transportation. *Padanna* rowers are forced to engage in other occupations like driving, sand mining, other agricultural and petty works.

Conflict and Survival

The river and the paddy field were considered as common property for fishing. When any conflicts arose about the fishing, the problem was solved by the fishermen co-operative societies. The prime consequences of the conflict–destruction of resource and marginalization of those who labour—are, therefore, central to the logic of profit-making. Now-a-days, the paddy fields are being given for non-*Padanna* fish farming by owners who are blocking the natural fishing and depositing farming varieties of fishes by destroying the natural species and polluting the water. Thus, *Padanna* people and other village communities have been expelled from their traditionally enjoyed natural resources and they are facing poverty, unemployment, and marginalization. Therefore, in some places unorganized resistances and protests have come up against it.

Conclusion

At present the habitat of the community is vastly changed. Construction of dams and bunds has altered the ecological pattern. Earlier, flood was a regular feature in the area and it deposited fertile silt in the paddy field and terrestrial land. Flood was essential for maintaining the fertility of the ecosystem. Along with this, it also enhanced the reproduction of the fish varieties. Excessive sand mining resulted in the increase of the depth of the river, and this has created hurdles in fishing. Moreover, earlier, during the summer seasons, the water level in the river used to get reduced, which supported fishing during the summer and subsequent monsoon season. Environmental imbalances have reduced the fish yield. Enhanced means of transportation, have also wiped out the inland navigation of the community. Moreover the profit-making fish farming in the paddy fields has ruined the natural fish varieties and *Padanna* people's rights over the natural resources.

12
Subsistence Pattern and Changes in the Ecology

A Study on the *Mavilan* Community of Kasaragod District, Kerala

K.P. Suresh

Introduction

Mavilan is an ethnic community inhabiting the Kannur and Kasaragod districts of Kerala. In earlier days, *Mavilan* were dependent entirely on the forest and forest resources. In those days, their main subsistence activities were food collection and slash and burn cultivation. During a later period, when a few landlords captured the forest land and imprisoned the *Mavilan*, they began to be treated as agricultural serfs. During this period too, the *Mavilan* could engage in their traditional subsistence activities as the landlords allowed them to follow their traditional mode of living. Since then, there was an influx of settlers that brought about major changes to the ecology and the subsistence activities of the community until then. It is these changes that are discussed in this chapter.

Area and People

According to the 2001 Census, the population of Mavilan in Kasaragod district was 25,110. Out of this, 12,310 were males, and 12,800 females; in Kannur district alone, their population was 4,403, out of which 2,200 were males and 2,203 females. In Kasaragod district, their settlements are in Chengala Puthigae, Karadukka, Kodom, Belloor, Kinnanur, Karimthalam, Badiyadukka, Balal and Panathady Panchayats. In Kannur district their settlements are seen in Peringome, Kakotalapeaddappu, Alakode, Udayagiri, Sreekandapuram, Payyavoor and Enivassery Panchayats.

Method of Study

Present study is based on both primary and secondary data. Fieldwork was conducted among the Mavilan settlements of Kasaragod district. For the purposes of the study, four Panchayats were selected from the two Block Panchayats, viz., Kasaragod and Kanhangad. Kuttikol, Bedadukka, Kallar and Kodom-Belur were the Panchayats selected for the study. An interview guide was used for collecting the ethnographic data of the community. Indepth interviews were used for collecting the details of the community. Secondary data were collected from different sources such as Panchayat Vikasana Rekha (Development Report), and published books and journals.

Subsistence Pattern

A society uses different methods to obtain its food and other necessities (e.g., foraging, pastoralism, horticulture and intensive agriculture). The *Mavilan's* subsistence pattern is at one and the same time, food gathering, hunting, fishing and agriculture (*kumeru* slash and burn cultivation). From the present food collecting methods and techniques, we can infer that the *Mavilan* were traditionally hunters and food gatherers.

Mavilan collected tubers from the month of *Dhanu* (November-December). The tubers are categorised as fibrous, non-fibrous, poisonous, and starchy. They used *natta* (digging stick) for removing the mud. The starchy tubers are *kurundu, mani kurundu, venni, muduka, pallathre,* etc. The tubers of *kurundu* are climbers, very small and growing just below the surface level. These tubers are identifiable from their leaves. The *venni,* and *muduka* (*Diascora* spp.), very big tubers, grow deep. For extracting these tubers they dig eight feet away from the stem base, because, these tubers are elongated and grow more than a metre deep. So they cannot extract out these tubers fully. *Ketha* is a poisonous tuber, but they consume it after detoxifiying processes. Of the *nara* and *chava* fibrous tubers, *nara* is an elongated tuber located by tapping over the mud and identifying their presence from the particular sound. *Mavilan* categorised the *nara* into two varieties, the red and white types, based on the colour of the tuber.

Honey Collection

Mavilan identify four kinds of honey such as *paithen* (*perumthen*), *cheruthen, kolethen* and *puttuthen*, named after the bees and the types of their hives, they find in the forest.

Perumthen: The bee that produces this type of honey is *Apis dorsata*. This type of honey is seen on the branches of big trees. Three to four pots of honey can be collected from it. Honey collection was practised at night. Bees of *perumthen* are dangerous, and, therefore, for collecting this type of honey they used *chuttu* (fire with sticks). Three or four men together used to gather and go for collecting this honey. One of them climbed up the tree while the second man stood aside with fire. They set fire to the hive, killing or driving away the bees. They then cut out each comb of honey and put in pots. The honey was then brought down by using string or fibres and they ate the half burnt bees.

Cheruthen: The bees that produce *Cheruthen* are *Trigona* spp. and these are the oldest and the smallest bees in the world. Although these bees do not sting, they bite. These bees are found in the hollows of trees and rocks. They store the honey in wax bags. Only a small amount of honey can be collected from these. The community members recognise the medicinal value of this type of honey and use them for such purposes.

***Kolethen*:** The bee is *Apis florea*. This kind of honey can be collected easily. It is a single comb hive. The hive is seen on the small branches of trees. The bees stay in a particular place for a certain period only. *Mavilan* wait for the bees to fly away and then drink the honey without leaving any trace.

Puttuthen: This is also called *thoduthen*. The bee is *Apis indica*. They are bigger than the *cheruthen* bee and stay in the same place and live for more than a year. The hive is made in the hollow of trees or inside the rocks. These bees make more than three honey-combs. To avoid the stinging of bees, the *Mavilan* chew *vatta* (*Macaranga indicalish*) leaves and blow the honey comb and make the bees fly away.

Hunting

Mavilan have ceased to be food gatherers and hunters, and have taken to a settled life (Luiz, 1962), but to appreciate the diversity of their traditional knowledge and awareness of their ecosystem, it is worth documenting what now remains in memory. Their hunting technologies depended upon the animals to be caught. *Mavilans* mainly hunted *malan* (sambar—*Cervus unicolor*), and *kattupanni* (wild boar—*Sus scrofa*); Their hunting techniques can be classified as (i) Hunting with traps and snares; and (ii) Hunting with the help of tamed animals. Their hunting mainly depended on trap and snares, but they also used *pirivu* (bow and stone), and also tamed animals.

Pirivu: It is a bow type transfixing trap to catch the birds and small animals. For making *pirivu* they use bamboo, cut in about three feet long and treated. The dried outer layer of *koli* (a climber plant) is made into a smooth fibre and this rope is tied on the bamboo stick in two layers, so as to make a *thattu* (a square network on the rope) to enable release of the stone when needed. At the top of the layer they put a *kenimkole* (a pointed piece of bamboo). This enables the stretching of the layer. When they see the game, they take the *pirivu* on the left hand and take stone on the right hand for aiming. They carefully select the shape of the stone to help in better aiming.

Adichil and Anchal: *Adichil* and *anchal* are deadfall traps. These consist of stockades on either side of the entrance of a fixed area. This stockade is to support a heavy crosspiece at the top. A bait is attached to the crosspiece in such a way that a slight pull to the bait string will result in the sudden collapse of the cross beam upon the prey, thus injuring the animals and capturing them inside the fenced arena. The *Mavilans* captured animals and birds including wild boar and sambar. In the case of *adichil* they used big stones instead of a cross beam.

Parakuzhi: It is a kind of pit trap. The technique of *parakuzhi* is very simple. They make a pit on the path that game frequents, and then fix some sharpened bamboo poles inside the pit. The pit is covered with sharp spikes or leaves. When the game falls into the pit the bamboo pierces its body. The *Mavilans* used to trap boar and deer using this technique.

Udumbu Piditham: In this game hunt, the *Mavilans* chase the *udumbu* (monitor lizard—*Varanus* spp.) and drive it to the nearest tree and then capture it by using a snare. This snare is made of *otta* (reed) and attached to a *thoti* (pole). They trap the *udumbu* when it falls on the ground and then beat it to death.

Mudikeni (spring trap): This trap is mainly for capturing *kaattukozhi* (wild fowl—*Gallus sonnerali*). A small oval ground is generally selected which is guarded by wooden fencing. A tree branch is bent and held down with an attached noose set at the entrance of fenced area. When bird enters the noose, it struggles to escape and this dislodges the weight, causing the tree branch to snap up into the original position, trapping the fowl.

Spring Pole Trap: This hunting trap consists usually of a noose attached to a tree branch bent down forcibly. When hare (*Lepus negricollis*) or other small animals touch the bait out of curiosity or accidentally, the trigger is set off and the force sends the branch back into its normal position, closing the noose upon the animal and hoisting it into the air.

Mavilans were clever in hunting *kattupanni* (wild boar), *malan, eyyan* (Porcupine—*Hystrix indica*), *kalamaan* (antler), *muchhan* (monkey), etc. Before the hunting they surround the forest with trained dogs. The dogs chase the animals and they spear them down (Madhavan, 1984). As hunting is now prohibited by the State, they have had to give up these livelihood practices.

Fishing

The fishing techniques of *Mavilans* include trapping and poisoning. The fish traps are mainly two types: *meenkudi* and *embakottae*. The simplest type of trap is *korattae*; it is a curved basket made of reed or bamboo. This is used for scooping the fishes from *thodu* (stream).

Generally they caught fish from the stream very near to their habitat. In the month of *Karkkidakam* (July-August) the *kammaue* fish swim from the river to the stream for laying eggs. Aware of the movement, pattern of the fishes, they place the trap on the side of the stream. When fishes come out of the river and swim into the stream they enter into the trap basket. Once they enter the trap they cannot escape.

Emba kottae is used in the rainy season. It is a basket trap; it is placed at the mouth of the stream. A *Mavilan* makes a platform at the mouth of the stream and onto it, this trap is attached with strings or ropes in such a way that when fishes leap from the platform they fall directly into the basket. *Kammaue* and *mudu* fishes are caught using this technique. In summer season they used knife to catch fishes. Especially the fish called *poyya kurudan* is knifed down. They caught *njandu* (crab) with bare hands, sometimes also using stones.

Fishing by Poisoning

During summers, they catch fish by poisoning. This technique is called *nanchidal* and this is a river-fishing practice. They prepare a poisonous mixture by using *nanchu kai* (*Croton klozschianus*, a poison fruit), *parankile* (red chilli) the bark of *vaka* (*Sterculia villosa*) and mud are ground into a paste. Making a bund in the river they change the direction of the water and mix the poison into the water thus impounded. The stupefied fishes that cannot swim are then caught by the *Mavilan*, using *korattae* or scooped up with bare hands.

Fishing is a group activity for the *Mavilans*. Their fishing groups include women and children. In summer season, they go for fishing as groups. More than one family is engaged in fishing. Mostly people from one settlement are collectively involved in fishing.

Shifting Cultivation

It is a type of prime cultivation, now extinct, generally done in the forest tracks for two or three consecutive years after which the land is left fallow as the soil loses fertility. This cultivation is a slash and burn type cultivation followed by most tribal people of India.

Kumeri Cultivation

Punam and *kumree* are varieties of shifting (slash and burn) cultivation practiced in the Malabar and the South Canara districts, respectively. The practice is called shifting cultivation because a site is cultivated only once and a new site will be slashed and burned every year. The frequency of a new site being cleared again is usually 12 years (Amruth, 2004:23).

In the month of *Dhanu* (December to January) *Kiragan* (headman) and a group of people collectively decided on a *kunnu* (hill) to be cultivated. Before cutting trees, all of them pray to their hill god for better harvest. *Kiragan* offers tender coconut to the god. After that *kiragan* selects a *kotta chedi* (*Grewia nervosa*), a shrub found in the hill tracks of their settlement, and they clear all around the chosen plant. *Kiragan* then cuts a coconut and pours its water around the plant. Then he cuts the small branches of the plant in three strokes with a knife. The cut off branches are thrown to the next hill. The branch is called *kottakole*. They believe that, if *kottakole* (stem of the *kotta* plant) fell in the east-west direction, the yield would be good. They repeat the action 3 or 4 times, if it does not fall in the right direction at the first throw.

Cutting of the trees and vegetation might go on until *Makaram* (a Malayalam month between January to February), slashed vegetations are left in the sunlight for several days. After one month, the boundary of the plot is covered with branches. It is done to control the fire from spreading over, and is called *beeri*. The headman sets fire in the slashed area of the valley. Later they remove the stumps and other unwanted wooden parts from the plots. Then they till the soil with *paruvae* (an implement used for *kumeri* cultivation). At the onset of monsoon, they sow seeds of different crops like *paru* (paddy, *Orizya* spp), *muthari* (ragi, *Eleusine coracana*), *thina*, *cholam* (maize), and seeds of vegetables, pulses and oil seeds. *Kiragan* takes a handful of paddy and throws it to the field on the day of sowing. Sowing is a collective activity; all members take part in the operations. Womenfolk scratch the top soil with *paruvae* and then sow the seeds and cover it with soil.

Diversity of Crops

They used to cultivate different varieties of paddy. Some of the elderly *Mavilans* say that they had 101 varieties of paddy they once cultivated and recall the names of some important varieties like *chennellu, kaliyan, kayema, kunhutti, thadichil, karinthadichil, vellutharan, thonnuran, raven, paliyan, kallae kottae, ponnarian, maleyedubban, okapadi, chathappan, kaipan, chemmolu*, and *koyivalan*. Some of the paddy seeds have names appropriate to its characteristics. For instance, *kunhutti* has very small seeds and *kaipan* is black in colour and bitter in taste. They also cultivate ragi, *kadu* (mustard), sown in the *mappa* (cutting bushes and branches deposited in a corner of the field). Maize is sown in the borders of the paddy field and the vegetables are cultivated in different corners of the field.

To protect crops from diseases, they used certain herbal medicines. They identified different types of diseases, for example *neeru vembu* (a type of disease) which affects the crops through water. Most of the crops had the power to resist diseases, but rarely though, insects attacked the paddy. Protecting crops from wild animals was another major task. Wild animals and pests like sambar, bandicoots, rabbits, rats, etc. caused severe damage to crops (Suresh, 2003).

Harvest begins in the month of *Kanni* (in the month of September-October). Before starting the harvest they make a *kalam* (yard) in order to thresh the paddy. They begin the harvest with the cutting of three *kathirs* (ears of corn), from these *kathirs* the *kiragan* takes one with *polivalli* (a fern species), and wraps it in *punna* (*Calophyllum*), Indian laurel leaf, and ties it to the *kundam* (pole). It is fixed in the centre of the threshing yard. Both men and women take part in the harvest. The reaped *katta* (the sheaf of corn) is brought to the previously prepared *kalam* for threshing. In olden times, when *Jenmi* (feudal lord) used both cattle and people for threshing paddy, the *kundam* was used for tying cattle.

Paddy, ragi, *thina, thora,* and other grains are dried properly and kept in bundles called *mooda,* made up of straw. The seeds for sowing are also hung above the hearth in *mooda*. Vegetable seeds for sowing are immersed in cow dung paste and stuck to the mud walls. Sometimes, they mix the seeds with ash before pasting on the walls as both ash and cow dung is believed to have properties to protect the seeds from attack of microorganisms.

Eco-cultural Changes with the Arrival of the Settlers

"Malabar presented a picture quite different from those areas in the region for a variety of reasons. The most dominant among them was the administrative and land control privileges of the *Jenmis* or landlords. The landlords dictated the land-use and fate of the forest for more than a century. They were slow to expand lands for cultivation; they also prevented their tenants from moving into the forests" (Amruth, 2004:16).

During the Second World War, when there was scarcity of food and lack of jobs for subsistence, from some Parishes of the Kottayam Diocese, people started to migrate to the Malabar region. As part of this migration, they bought 1,800 acres of land from the Nileshwaram *kovilakam* (royal family). The migration to Hosdurg area started with the 1950s. Encroachment and evictions presented often raucous situations in the social/political spheres of the State during the 1950s and 1960s (Amruth, 2004:17)

During 1959-65 the migration was under the leadership of the Nair Service Society (N.S.S). They bought 500 acres of land from the Nileshwaram Raja. In 1970, forty families came to Madathumala (migrants now call it Ranipuram). The migration was at its peak during the 1965-75 periods. Until the 1970s, the *Mavilan* practised *kumeri* cultivation. After the Land Reforms Act was passed, many of the *jenmi's* lands went into the hands of migrants. The successors of the *jenmi* also transferred to migrants some of the lands in which *Mavilan* used to cultivate. The *Mavilan* who had no landholding certificates, had to shift to other areas. In Panathady, besides *kumeri* cultivation, pepper, coconut, cashew nut, mango, jack, etc., were also being cultivated, while the migration and subsequent events that affected the life and livelihood of *Mavilan* occurred. The migrants began to cultivate tapioca, rubber, coconut, arecanuts, cashew nuts etc. widely and they began to concentrate mainly on cash crop cultivation.

Mavilan used to consume paddy, ragi, and vegetables that were cultivated in the *kumeri*. When their cultivable land was lost with settlers coming, they were forced to live with the resources in the limited land available to them. During *jenmi's* time the *Mavilan* were only semi-dependent on the outside and the market; they depended on outsiders only for *uppu* (salt), *chappu* (tobacco) *unakku* (dry fish) and *thuni* (clothes). When the migrants shifted their cultivation from food crops to cash crops, the *Mavilan* were forced to change their food habits. They changed their traditional habits and they began to buy everything from market.

The *Mavilan* had possessed some land in Koyichtta, Parakadavu, and Prantharkavu. In these lands they cultivate paddy, vegetable, pepper and so on. In harvests they used to get sufficient yield. When settlers changed the land-use pattern and started cash crop cultivation, they began to use high yielding varieties and chemical fertilizers for more profit. To combat the increased plant diseases and pests, they used more pesticides and fungicides. The *Mavilan* still followed their traditional farming system, but their traditional techniques were insufficient to control pests. So *Mavilan* and other tribal people stopped their own agriculture and began to work as wage labourers for the settlers and other non-tribal people who needed help in their expanding cultivation.

In the past *Mavilan* used their herbal medicine for all their diseases. But due to the interaction with the non-tribal people and modern education, the youngsters have lost faith in their traditional medicine. They now go for modern medicine for common diseases. The old tribal healers who prefer to use their own medicine for all diseases also believe that their traditional medicinal treatment was not enough for the new diseases for which they too depend on the primary health centres. This also adds to their neglect of the bio-diversity in their ecosystem.

Conclusion

Earlier *Mavilan* were food gatherers and hunters and they depended on the forest for their livelihood. In the changed circumstances of land and livelihood alienation, they are fast losing their traditional knowledge of their predecessors regarding the rich diversity of their ecology as also the diversity of their eco-system. Therefore this chapter has attempted documentation of their knowledge, techniques and livelihood practices that were based on such intimate knowledge of their ecosystem. This chapter also attempted to show how these changes happened in this community over the years.

REFERENCES

Amruth, M. (2004). *Forest Linkage and its Implications on Forest Management: A Study of Delampady Panchayat, Kasaragod District, Kerala,* Discussion Paper No. 69, KRPLLD, Thiruvananthapuram.

Census 2001.

Luiz, A.A.D. (1962). *The Tribes of Kerala,* New Delhi: Adhimajathi Sevak Sangh.

Madhavan, M. (1984). *Velicham Pakaru,* Kanhangad: Janatha Printing and Publishing.

SamagraVikasana Rekha, (1996). Panathady Panchayat, Janakiya-Suthranam.

Suresh K.P. (2003). "A Traditional Community in Transition: A Study of *Mavilan* in Kasaragod District of Kerala", KRPLLD Project Report (unpublished), Centre for Development Studies, Thiruvananthapuram.

13

Impact of Social and Environmental Factors on the Health of the *Villiyaars* of Puducherry Union Territory

V. Pragathi

Introduction

There are three main factors which determine how healthy a person will be: genetics, environment and lifestyle (Smith, 1992). It is helpful to think of these three things separately although they interact and are not independent on each other. Genetic endowment cannot be changed but how the individual is affected by their genes will, to a large degree, be influenced by the way they live their life. It is convenient to divide living conditions into environment and lifestyle. Environment refers to aspects of a person's surroundings such as their housing, their neighbourhood, the water they drink and the air they breathe. In general, these are things over which individuals have little control. Lifestyle refers to aspects of the way people live over which they have some control, such as smoking, eating and drinking. The environment has a major influence on lifestyle and the lifestyle can modify the effect of the environment.

It has been noted by many that tribals living in remote areas have a better health status and more balanced food than those living in less remote areas (Chaudhuri, 1986; Chaudhuri *et al.*, 1989). Some of the anthropologists have also noted that physical well-being of the tribals is inextricably linked up with the forest ecology and rapid disappearance of forest, commercialization of minor forest produce, replacement of food crops by cash crops and tendency towards monoculture forestry and lastly privatization of communal lands may affect the health and nutrition and physical well-being of the tribal population.

Each of these aspects has a deep influence on health and to raise health status and quality of life of people. Sustained efforts are required to bring about transformation of the society as a whole in graded amplitude. Some attempts have been made to study the inter-relationship between socio-cultural practices, environmental conditions and health aspects among the *Villiyaars* of Puducherry Union Territory.

It is hoped that this health oriented action research study will help in the formulation of need-based health care strategy for the *Villiyaars* in particular, and other tribes of Puducherry, in general.

Villiyaars are not natives of Puducherry. Referred to as *Irular, Pucali, Kattukaran, Vettakkaran, Villiyaar* and *Yenathi-Villiyaar* by other communities, they are migrants from different parts of Tamil Nadu who have settled in Puducherry since long. They do not remember the period when their forefathers migrated and from which part of Tamil Nadu they came to Puducherry. However, some could recollect later migrations especially from *Agaram, Gingi, Bhuvanagiri, Thiruvannamalai, Irayur, Mathur* and *Omandhur* villages of South Arcot districts of Tamil Nadu.

The *Villiyaars* are traditionally catchers of snakes, rats and mongoose, who lived by selling the snake skins and mongoose fur, honey collecting, and fishing. Migration to Pondicherry and living on plains have made them move away from their traditional occupations and opt for wage labour in agriculture, brick kilns, wood cutting, etc., and herding livestock as main occupations although on occasions they still collect honey and catch snakes and rats from paddy fields. The Villiyaars, as per the Constitution of India, are Scheduled Castes in Pondicherry, but Scheduled Tribes in Tamil Nadu and are claiming ST status in Tamil Nadu also.

This chapter attempts to study the impact of social and environmental factors on the health of the Villiyaars of Puducherry district of the Puducherry Union Territory. It is based on insights from an ethnographic field study on the Villiyaars and an empirical group study using conventional anthropological methods. The study undertook (a) a general survey of their life cycle rituals and utilization of health services by them and examined the records of hospital, PHCs and Sub-centres, and other available secondary sources of information like District Gazetteers, Census reports, Statistical handbook, and special studies, reports and publication on the community. There is very little empirical and statistical information relating to the health care of the Villiyaars and on their environment.

Physical Environment

The present habitation of *Villiyaar* is a plains region which has a hot tropical, maritime climate, with high humidity and moderate rainfall. There is even scarcity of drinking water in the colony. Their stay in this open, dry land and dusty weather has brought them several respiratory and lung diseases and among them tuberculosis is found to be rampant. Added to these are intestinal diseases and malnutrition. There are no health and medical facilities either in the colony or in the vicinity.

Housing of the *Villiyaars*

Warmth and shelter are among the most basic needs of humans. People living in damp housing are much more likely to develop asthma and respiratory infections. This is particularly true for children. Crowded housing increases the likelihood of transmission of infection between residents. Houses of *Villiyaars* are constructed of stones and mud. Blocks of irregular stones are placed one above the other and plastered with mud. The houses have thatched roofs. These houses are generally rectangular. There are no windows in the houses. Houses of *Villiyaars* are filled with darkness even during day time. When meals are prepared the room fills with smoke and causes suffocation. It has been observed that the conditions under which *Villiyaars* live have a major effect on their health status.

Houses of *Villiyaars* do not have any arrangements for light and air. Poor lighting during night hours, absence of suitable exit for smoke or too much smoke in crowded and ill-ventilated houses may be the cause for the high incidence of night blindness found among them. Lack of fresh air during night causes respiratory troubles. Traditional customs, ignorance, illiteracy and poverty on the one hand, and indifference and reluctance of the management in these regards on the other hand, are the main causes for their living in this condition.

Hygiene and Sanitation among the *Villiyaars*

Sanitation among the *Villiyaars* is very poor. As a result, they are now facing a lot of problems with regard to their health and sanitation. The open space close to their colony is used for answering nature's calls. Young and the very old use the space immediately around the houses for the purpose. Added to this are the cow-dung and other filthy matter which get collected in the manure pits located near the houses. Rain and bath water gets collected in them and this becomes a breeding ground for mosquitoes, flies and other harmful insects. Existing habits of defecation are largely responsible for the ill-health of the *Villiyaars* in these areas. Medical doctors at thePHC mentioned that the germs of cholera, typhoid and dysentery are present in the stool of affected men. During rainy seasons, these germs get mixed with stream water and this contaminated water is used by the *Villiyaars* and thus they suffer from these diseases. These parasitic infections can be controlled by providing sanitary facilities to the *Villiyaars*.

Personal hygiene is also quite poor among the *Villiyaars*. This is due to poor awareness about the importance of personal hygiene. The State of environmental and personal hygiene was found very deplorable among them. Due to poverty and illiteracy the *Villiyaars* are unaware and ignorant about the fatal consequences of such poor environmental sanitation.

Drinking Water Supply to *Villiyaars*

Many infectious diseases such as cholera, typhoid and polio are carried through water. The organisms that cause these diseases can get into drinking water if human excreta are not properly disposed of. It is needless to emphasize the necessity of developing safe and hygienic drinking water in each and every village of the tribal areas, where high infestation in almost every village results in diverse health problems, untold misery and suffering and loss of considerable manpower. The principal source of these infections is the polluted drinking water from village ponds, rivers, hand pumps, tanks, etc. On account of the rocky terrain and the lack of facilities for hygienic drinking water, people have to fall back upon the rain water collected in ponds, rivers wells and hand pumps. Water is taken directly from rivers, wells and hand pumps for consumption. Water crisis in the interior villages becomes acute during summer.

Food Habits of *Villiyaars*

In the past, food habits of the *Villiyaars* were dependent on various forest produce and flesh of different animals. Now their staple food is rice. They take two meals a day, one in the morning and the other at night. Generally they purchase rice from ration shops run by the Government. Occasionaly they still visit the forest and collect various tubers, roots, fruits, honey, mushrooms, etc. Now the *Villiyaar* eat *kambu, cholam* and *kelvaragu*. The food taken by *Villiyaars* is neither nutritious nor does it meet the caloric, protein and vitamin requirements of the body. Milk is only consumed occasionally. The consumption of cereals, pulses and vegetables is very less, mainly because of low production and cost factors.

Smoking and Drinking Habits of *Villiyaars*

The *Villiyaars* are fond of smoking and drinking. It has been observed that the majority of informants smoke (93.3%) and drink (86.0%). They use tobacco and locally available leaves for smoking in the form of bidi, and locally prepared alcoholic drink is very popular. There is no restriction on females either for smoking or drinking along with their spouses.

Factors responsible for Poor Health Status of *Villiyaars*

The poor health among the *Villiyaars* is due to lack of nutritious food, unhygienic living conditions and smoking and drinking habits and other environmental conditions. Smoking and drinking habits result in malnourishment, loss of body weight and decreased resistance to infections. Unhygienic living conditions, delayed treatment of diseases and malnutrition affect their health badly. Even pregnant and lactating mothers and children are not provided with nutritious diet. Several diseases are caused due to lack of safe drinking water. The *Villiyaars* as a whole do not attach much importance to sanitation aspects, be it sanitary latrines or systematic disposal of waste products.

There is paucity of comprehensive health related studies among the tribal population. Most of the existing studies are isolated and fragmentary in nature. There is urgent need for initiating area-specific and group-specific, health-needs action research in consonance with the felt needs of the tribal communities. In India, studies of health behaviour, treatment process and health culture of the tribal people are of recent origin. It is quite well-known that in India there exist not one, but a plurality of health care systems and norms, beliefs and practices regarding health and treatment process. The folk medicine is one side of the spectrum, while on the other side, there exists modern scientific medicine system. Thus in contemporary India there exist plurality of health cultures and numerous health care systems. The health culture of tribals has attracted the attention of anthropologists. However, the studies conducted by anthropologists on tribal health emphasize, rather over-emphasize, the cultural dimensions of health. Moreover, these studies suffer from the lacuna of treating the cultural factors in isolation. Cultural factors reveal only a partial truth. They can explain to some extent the health culture of tribals. But this is not all, because many factors interact and impinge upon health. Health is as much a cultural phenomenon as it is a reflection of the political, economic, environmental and social forces. Health is very intimately connected with ecology and culture. Therefore, this study will help in understanding the health problems and health care system of the community under study, using holistic and interactive approach. In a nutshell, the observed levels of living and health status of the *Villiyaars* is the result of several associated factors, viz., environment, economic, social and dietary, that have caused or contributed to widespread malnutrition and associate deficiency diseases.

REFERENCES

Chaudhuri, B. (1986). Social and Cultural Aspects of Health. *Journal of Social and Economic Studies*. 39: 379-388.

Chaudhari, Buddhadeb, D. Dasgupta and K.K. Chatterjee, (1989). (Ed.)

Tribal Medicine. Midnapore. West Bengal: Regional Research and Study Centre.

Smith, A. (1992). Setting as Strategy for Health. *British Medical Journal*. 304:378-398.

14
People's Struggle and Adaptation with Changing Environmental Phenomena in Sunderbans Delta Regions

Mohan Kumar Bera

Introduction

'Sunderbans' literally means beautiful forest. Beautiful, indeed it is. There are two legends about the origin of the name. One is that it is named after the dominant tree of the region, 'Sundari' (*Heretiara litoralis*). The other says that the name has refernce to its nearness to the sea or 'Samunder' in the local dialect. So, a forest or '*Ban*' near the sea or Samunder became Samunderbans or the Sunderbans (Chathopadhya, 1995; Zockler, *et al.*, 2005). The Sunderbans has a truly unique ecosystem. Apart from providing home to a number of rare and endangered flora and fauna, it is the only mangrove forest in the world inhabited by tigers (Jalais, 2007).

The history of the human settlement had begun some five hundred years ago. Khan Jahn passed a *Jagir* for Sunderbans in 1459 (Chottopadhya, 1995:3). But actual land reclamation started largely between 1765 and 1770 with the patronage of Collector General, Claud Russell. Mr. Henkel from present Hingalganj attempted to clear 7,000 acres in 1982 (Jalais, 2007:336; Chottopadhya, 1995; Bera *et al.*, 2006:6). The Colonial Government wanting to convert this mangrove land into safe and productive agricultural field distributed it to *jamindar* for a long term lease. Mainly socially and economically marginalized people from Midnapur district, Bankura, Purulia, Howrah and neighbouring States came as labourer for clearing the forest and to make embankment and cultivate. Many of these labourers left after cultivation, but some of them stayed permanently. Now the total population is 4,20,000 (Basu, 2006), spread over eighteen Blocks.

As per the UNEP (2007) report, Sunderbans delta region is most prone to flood due to increasing levels of tide. Research team working for the Government of India first noticed in 2001 that the islands are vanishing. The team projected only 100 islands though it was 102 (UNFCC, 2001). Prof. Sugata Hazara predicted that more than 70,000 will be environmental migrants in next twenty-five years (*Telegraph*, 30th October 2006; CSE, 2007). UN accepted that there is a close relation between climate change and hydrological disaster (IPCC, 2001; ISRD, 2002, 2004; Twing, 2004, Marfai *et al.*, 2007), causing 'vanishing islands'.

Though there are many incidents of breach of the embankments during high tide (*jowar*) every year, people are still staying in these islands. They have adapted themselves to the crisis and have developed alternative supporting livelihood system to deal with the changing climate and the floods. UN-Hyogo Conference (2005) also highlighted the importance of community adaptation to climate change.

Being a very remote delta region, many island villages in the Sunderbans region are beyond research. The impact of rising sea level and disaster is more in the southern and the south-eastern parts of Sunderbans which cover Pathar Pratima Block, Kakdwip, Namkhana, Sagar Block (Guha 2006; *Telegraph*, 30th October, 2006). Agricultural land, houses, pond, playground, and tube wells are being captured by estuaries, but people are struggling with the changing environment, a challenge that was not there some twenty-five years back.

The destruction of the Sunderban region had started in the 16th century by the colonial government (Desuza, 2001; Rudra, 2001; Kanjilal, 2000). Added to this, the natural ecosystem of mangrove has been converted into paddy field by the migrants from East Pakistan, neighbouring districts and States. Every household does not have the capacity to purchase a square feet of land in another island. If they migrate to some other island, then there is a question of livelihood. So they developed certain understanding to deal with the rising sea level. In this context, this chapter tries to explore local people's perceptions of changing environmental phenomena, their skills of adaptation to deal with these phenomena, and the patterns of their adaptation to cope with climatic change and disaster.

Conceptualization of the Research Problem

The term 'adaptation' refers to the coping mechanisms that humans display in obtaining their wants or adjusting their lives to surrounding environment. Since early period, empirical studies have reflected that ecology plays an important role in human adaptations and maintenance of the culture and life style of a community. The geographical location of a place is mainly responsible for its people's social system, customs, behaviour, religious beliefs and tradition and mental make up and shaping their livelihoods. Different ecological settings and practices are linked to the use of available natural resources. It is believed that environmental condition of a particular geographical location determines a people's practice of distinct activities (Ratzel, 1899; Semple, 1911; Hussain, 1996; Basu, 2006).

The changing environmental condition influences the life, culture, interrelation and interdependency between man and his environment. It is also argued that the local technology, management strategies and local culture are interrelated with environmental condition of a particular region which helps adapt to natural phenomena (Basu, 2006; Sen, 2003).

Adaptation enhances the strength of a people to compromise the internal and external environmental changes (Sahlins, 1971). It is a highly systematic process which involves very high social and human costs. Adaptation to environmental degradation and disasters includes technological, institutional, economical and behavioural options which frame the broader developmental strategies and policies (Klein *et al.*, 2007).

When one society lives in any particular geographical region for generations, its people develop certain understandings and knowledge to deal with every phenomenon. Adaptation to a particular environment is the representation of their understanding. The villagers of environmentally vulnerable countries are known to change their cropping pattern, agricultural systems, fishing practices and other livelihood supports to cope with environmental change (Allen, 2006; Klein, 2007:147). For example, the Zimbabwean farmers change their agricultural practices as per the prediction of environment. In Bangladesh, geographically most vulnerable island (Char) communities are not found migrating in the post-flood period, but are found to have developed changing patterns of livelihood systems from their flood-affected agricultural land. So the changing livelihood systems enhance their capacity to deal with natural phenomena and reduce their vulnerability.

Adaptation to environmental changes depends on community capacity, community resilience and social system (Allen, 2006; Kundzewicz, 2002; Blaikie *et al.*, 1994). Adaptation capacity varies from community to community in different geographical locations. Adaptation strategies of certain communities may create mal-adaptations for certain other communities. In the Indian context, adaptation capacity and strategies vary in different communities belonging to different caste, class, religion and ethnic backgrounds.

As per Blaikie *et al.*, (1994) the socio-economic vulnerability is defined as the function of adaptive capacity of the society in a certain geographically vulnerable location. Adaptive capacity is the function of social and economic processes. The Pressure and Release model of risk identified the factors which affect the community's adaptation capacity to changing environment. According to Blaikie *et al.*, (1994), unsafe condition arises from political, economic, social and institutional processes which directed the community adaptation capacities and strategies. Capturing of the land of the Sunderbans region had started during the colonial period which disturbed the natural ecosystem of this region (Chapman, 2007; Desuza, 2001; Kanjilal, 2000; Basu, 2006). The socio-economically marginalized people get livelihood support from this reclaimed land. So the people of Sunderban developed their distinct adaptation capacity to deal with any type of environmental problem.

As per report of United Nations Framework Convention on Climatic Change (UNFCC: 1997) and IPCC (2001), 'adaptation' is the adjustment in natural or human systems in response to actual or expected climatic stimuli or their effects, which moderates harm or exploits beneficial opportunities. Natural phenomena are not

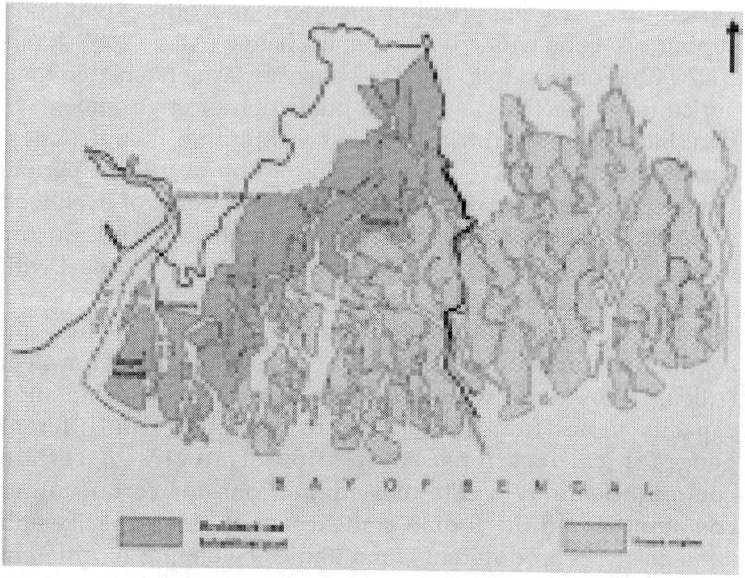

Fig. 14.1: The Sunderbans forest region is demarcated by *Dampier-Hodges line* between India and Bangladesh (Gopal and Chauhan, 2006).

controllable. So increasing the adaptive capacity is best measure to reduce the impact of climatic change (UNESCO, 2003). The Millennium Development Goal (2005-15), IFRCC (2004), ISDR (2005) also have given importance to reducing the losses due to environmental disasters by enhancing the community adaptation capacity and direct involvement with their knowledge.

Research Site

The Sunderbans is spread over an area of 9,630 square kilometres in the district of North 24 Parganas and South 24 Parganas of West Bengal, India. Sunderbans region is located between 21° 31′ to 22° 38′ N Lat. and 88° 5′-90° 28′ E Long. covering a forest area of 4,284 square kilometre which comprises 102 deltas. Presently 54 deltas are inhabited by people, which cover 4,284 square kilometres (Bera *et al.*, 2006). Sunderban deltas were formed by the deposited sediments from three major rivers, i.e., the Ganga, Brahmaputra, and Meghna and their dense network of smaller rivers and channels. The maximum elevation is only 10 m above the mean sea level. The western limits of the Indian Sunderban are demarcated by the Hoogly river and the eastern boundary by the Raimangal river.

Geology

By the geologist's scale, land formation of this region is a recent phenomenon. The alluvial condition is hardly conducive to fossilization of animals and plants. There were several geomorphologic changes that resulted in deposition of sediment in the Bengal basin and developed the Sunderbans Delta (Wadia, 1961; Gopal and Chauhan, 2006; Chottopadhya, 1995; Banerjee, 1998). During the 16th century, the Ganga river changed its course to shift eastwards and join the Brahmaputra river. The continuous plate tectonic activities influenced the hydrology of deltaic region because of changes in sedimentation patterns and reduction of fresh water flows. Most rivers other than Hooghly contributed to the formation of deltas and lost their original connections with Ganga (Gopal and Chauhan, 2006).

Climate

Climate of the Sunderbans region is characterized by high temperature and humidity (>80%) through the year and well distributed rainfall throughout the year by typical tropical monsoon climate. The maximum average temperature is 29°C that is felt during the month of June. But the highest temperatures cross 32°C during May-June. The average rainfall is 1,800 mm. Recent reports suggest that the air temperature over the Sunderbans Region and adjacent of the Bay of Bengal is gradually increasing (Gopal and Chauhan, 2006; Haq *et al.*, 1999; Agrawala *et al.*, 2003). The cold season covers mid-November to end February. The average minimum temperature of 20°C is recorded during December and January. Thunderstorms are common during the summer after monsoon. These thunderstorms are known as Kalbaisakhi or nor-westers. These thunderstorms result in heavy rainfalls and drop in temperature. The storms often developed into cyclones that raise the tidal wave up to 7.5 m high. There is correlation between rising temperature and flooding, erosion and saline water intrusion due to storm surges (Haq *et al.*, 1999;

Agrawala *et al.*, 2003; Gopal and Chauhan, 2006; Banerjee, 1998).

Soil Characteristics

Land is moulded predominantly by the tidal action of Bay of Bengal. Many creeks, network of estuaries, drain the land at each ebb tide. Rivers are straight, a consequence of high tidal forces. The easily eroded material from the river banks collect at the mouth of a river and form banks and char which are blown into dunes above the high-water mark by the strong south-west monsoon (Chauhan, 2006; Das and Raha 1963; Banerjee, 1998). Finer silts are washed out into the Bay of Bengal, but mudflats are also formed on the leeward side of dunes where they were protected from strong tidal forces.

Surface of this land is made up of alluvium with great depth. Surface soil close to the Hooghly is a hard clayey loam (*Entel*), while in the northern part, sandy loam (*Pali*) predominates and the eastern and central parts bear chiefly clayey loam. The alluvial soil in Sunderban area shows a white efflorescence of sodium chloride, as they are impregnated with salts by tidal estuaries. The soils are classified into saline, non-saline, and alkali soils. According to Das and Raha (1963) agricultural land can be divided on the basis of soil pattern into the following types:

(a) *Matial*: A clayey soil, whitish in colour, loose and light in composition. This soil is suitable for the '*Patna*' rice which is grown in the Sunderban region.
(b) *Baliara* or *Dosara*: The loamy soil, reddish in colour.
(c) *Dhap* or *Chura*: It is whitish in colour and higher level than other types.
(d) *Dhal*: It is reddish in colour and cracks when dry and not suitable for cultivation.

Demographic Structure

Initially, the Sunderbans was a no man's land. Mainly tribal and lower caste people from different parts of West Bengal and India were brought for clearing the forest. These inhabitants of the Sunderbans got their livelihood from the forest, rivers and reclaimed land. It is found that low-caste Hindus and Muslims resided in the North Sunderbans. Among the low caste Hindus, *Pods* were numerous in the west, while *Namasudra* and *Chandal* concentrated in the east, and *Mahishya* resided in south. Among the Muslims, *Fargis* are dominant (Chottopadhya, 1995; Basu, 2006).

In 1872, the total population was 2,96,045. But it increased tremendously in 1951 and 1971. Main reason was the massive migration during 1947 and the *Bangladesh Mukti Judho* of 1971 from East Pakistan. The people of Sunderbans are mainly dependent on agriculture and many villagers are also dependent on fishing, and honey collection.

Research Methodology and Data Collection

Community adaptation to environment is a combination of socio-economical, cultural, institutional, technological, and behavioural structure of a particular community. Adaptation varies from place to place though people belong to the same cultural groups and community. According to Tran (2007), the understanding of integrated activities of a community to deal with every natural phenomenon gives more insight in adaptation research, because adaptation is an everyday phenomenon and starts from each household which connects each and every member and entire community. To fulfil the research objectives which cover community coping mechanism, traditional belief and culture and strategies to dealing with changing livelihood, an ethnographic approach in qualitative research paradigm was used. Two villages were selected to conduct case studies to get more insight for the research. The main criteria for selecting the villages were :

(a) the villages which were more affected due to floods;
(b) the villages which were recently flooded and do not have any earlier experience of flood; and
(c) villages that have more migration and displacement due to natural disasters.

Two villages have been selected from Patharpratima Block named Purba Dwarkapur of Lakhsmijanardanpur Gram Panchyat, I-Plot and Dakshin Sitarampur of G-Plot Gram Panchyat. Purba

Dwarkapur village was badly affected by flood due to the breaching of the river embankment in 2005. Village was submerged in the river waters for seven days. Dakshin Sitarampur village is situated on the sea coast which is the most flood and disaster affected village of Patharpratima Block (Disaster Management Report, Patharpratima Block, 2008). Flood disaster is a recent phenomenon, started mainly after the '80s, and most parts of this village are already taken over by the sea. Many villagers have become landless labourers due to the sea encroachment and many of them have also migrated to other villages.

Primary data was collected during the field work in March-June, 2008. In-depth semi-structured interview has been conducted among the people of three age groups, i.e., above 65 years, 22-64 years and below 21 years. A sample of seven respondents from the first group (above 65 years), six respondents from second group (22-64 years) and two from the third group (below 21 years) were interviewed. Interviews were also conducted with the Block Relief Officer and a Gram Panchayat member. These interviews concentrated mainly on the community understanding of the climate change, its coping strategies to flood disasters and changing livelihood system and impact of the socio-cultural and historical changes on the community understanding.

Secondary data have been collected from Patharpratima Block Office and Gram Panchayat offices. Published books and journals, news reports, West Bengal government's unpublished report and disaster management reports were also studied very carefully and analysed.

Perceptions of Environmental Degradation

The perceptions of environmental degradation vary with the socio-economic structure to livelihood support and political enlightenment. The people's perception of the regularly flood affected region varies from that of the people of irregular flood affected region. Climatic degradation of a particular place is a kind of maladaptation for others. People from the non-flood affected region argued that the floods are created by river-side villagers to continue their wage labour. The whole perception of degrading environment can be divided into the following sub-headings:

Religious Perception

The people from rarely flooded region believe that climatic change and disasters are created by the gods. The gods cannot co-operate with sin and sinners. So, a disaster is the only weapon to destroy the sinners. The villagers believed that supernatural powers are behind the floods and disasters. The religious perception includes religious songs and myths, and folk tales (*Mangal Kabya*); astrological calculations and year books (*Full Panjika*); and, local religious activities and community *sangeet* (*Sankirtan*). As the earth is made up of land jungle and water, the *god of Floods* (Ganga) has good relationship with the *god of jungle* (*Banabibi*). *Banabibi* never likes floods or any other disaster. But '*Bahardar-Dakshin Rai* is involved with all the tragic activities in the Sunderbans. Goddess *Banabibi* supports and encourages the local people in dealing with all types of risks. The kernel of this story varies from place to place and community to community.

In the south-central part of the Sunderbans, people respect Goddess *Sitala,* who is a village goddess. The story of *Sitala Mangal Kabya* identifies the Goddess Sitala as rescuer and kind hearted to villagers. So, many times when there is a flood and any other natural disaster, people go to her temple and pray for their lives and properties, they play her story in an open theatre (*jatra*), and arrange *mangal sangeet*. When a troupe of *mangal sangeet* travels through the village, the villagers join in and greet them with playing *sankh* (conch). The *mangal sangeet* troupe uses *mridang,* and *karatal* and sing the *Sitala Mangal Sangeet*. Villagers put a boundary around the village with bamboo sticks and *neem* leaf. They also burn huge *Dhuna* (gum of *Sal* tree). The important story behind Goddess *Sitala* is that, one day, when she was taking rest after lunch, she heard a great commotion outside her temple. The goddess called her servant (*dashi/sevak*) to find out what was happening. *Dashi* climbed to the top of the temple and immediately cried out in panic because Goddess Ganga was coming at them with the entire force of destruction. Goddess Sitala came out of her temple and stopped Ganga, asking her why she was creating problem for her sons. To this, Goddess Ganga replied that she lost her nose-ring and was going to search for it. Goddess Sitala, in anger, threw her own nose-ring to Ganga and told her not to destroy the land, lives and property for her selfish interest. Goddess Ganga felt very sad. Before leaving the place she promised Sitala that she won't come again to destroy, but will develop the land instead.

The lessons that have come out from the stories are discussed as below:

(a) The floods and disasters in the Sunderbans are very uncertain. As a part of the river Ganga or Ganges and its channel networks, flood can happen any time. The disasters from floods could occur during high tide which breaches the embankments.
(b) Flood disaster not only destroys but also enriches the agricultural land which developed distinct alternative livelihood supports. Though there is climatic degradation, nature creates a support to deal with every type of environmental changes.
(c) Reduction in the impact of climatic degradation starts from every household. Group activities which support individual households' coping capacity are essential.
(d) People of Sunderbans have been dealing with floods ever since the fifteenth century. They believed that the playing of the *Mangal Snageet* is not only for religious purposes but also to organize the entire community to deal with natural phenomena and risks collectively. Social solidarity and communication have been set up by this activity.
(e) Use of the bamboo sticks and neem branches during *Mangal Sangeet* gives a very important indication–to manage the natural resources which are available in every household. For instance, bamboo has remarkable importance in embankment management and construction of houses in the Sunderban region. In the Sunderbans, when all fuel woods become wet, bamboo is the only support during post-flood times. Importance of *neem* is also highlighted ritualistically in this region. In the post-disaster situation, people use the medicinal *neem* leaves to prepare disinfected bathwater.

Scientific Perception

The scientific perception of environmental degradation, climate change and its impacts varies with educational level and experience. The government officials indicate the impacts of environmental degradation, in respect to the decreasing depth of the river and soil erosion. On the other hand, villagers identified the decreasing forest reserves, and the decreasing varieties of fish in rivers. The villagers of the sea coast area realized that there is an increasing force of sea wave which was not there twenty-five years back. The perception of environmental degradation also found through the increasing tide water level which the villagers experienced in Sunderbans due to increasing the height of embankment. According to the villagers, in the year 1943, the average height of the river embankment was just seven to eight feet, but today it is as high as seventeen feet. So it is easily calculated that in the last sixty years, height of the tide water has increased by around ten feet. But the people farther from the river embankments have very little experience about the climate change and its impacts.

Political-Ecological Perception

The environmental degradation in the region started in the colonial period itself. Villagers argued that the impact of environment degradation is exaggerated by the development processes in the Sunderbans. After the construction of the Farakka barrage, the magnitude of flood disaster has increased, and disturbs the natural ecosystem. People from the coastal area, South-East of Sunderbans realize that the decreasing forest cover in Bangladesh Sunderbans is affecting the West Bengal portion's natural ecosystem also. Tusar Kanjilal (2000) has raised the question: who degraded the natural ecosystem of Sunderbans?

Physical Adaptation

Adaptation strategy of the people in the Sunderbans varies from place to place. The preparedness to disaster and environmental degradation also depends on the economic strength and the social status. Those villagers, who lost their agricultural land, are still living near the river and sea coast. Adaptation capacity of the river side people is more remarkable than that of the people from the interior village. For interior villagers, the adaptation process of the river side people is one type of strategy to utilize the natural resources. But for the river side people, this is simply living with nature and no strategy.

The housing structure of the river side people varies from that of the interior part of the village. Most of the river side houses are earthen ones that are durable only for seven to eight years. When the villagers destroy the earlier structure of their houses, they never remove waste materials from the original place. After demolishing the earlier structure, they construct the new house over it so that the altitude of the houses automatically goes up. When the altitude is higher, there would be less intrusion of saline water (*nun jal*) and salty soil (*nona mati*).

The houses are built with the materials available in the village itself. The height of the river side houses is not more than twenty-five feet and many times these are huts (*tung pala*) only. Most of the houses have only one big room which is separated by small walls to be used for different purposes. As the villagers in the river and coastal sides, experience greater wind velocity, they never increase the height of the house, which could be most prone to the flood disaster.

The roof of the houses varies from household to household. Most of the river side houses have roofs made of paddy straw (*khar*). The main reason behind this type of roof is the weakness of the wall and the high wind velocity. They divide the total length of roof into two parts, i.e., top and lower roof. Sometimes the lower roof is made of tiles, but the top portion is made of straws only. Presently, many households are using tinplates for the roof. But the villagers opine that they are more comfortable in straw-thatched houses. Villagers prepared net with dry straw and sometimes nylon nets purchased from market and covered the entire roof to protect their roofing material, mainly straw, from the high velocity wind.

The walls of the houses are constructed layer after layer. Once the first layer is dried, then the next layer of mud is added. During the construction of an earthen wall, villagers first erect bamboo sticks and cover them with mud to make the wall stronger. Selection of mud is a very important task for the senior and experienced architects of the village. They come together and select the land. The architecture of the house is not decided by the household alone, but is a decision taken by a group of villagers.

Pattern of Land Utilization and Water Resource to Changing Environmental Phenomena

Agricultural land is the main livelihood support in the Sunderbans. Though, these are reclaimed lands from the seventeenth and eighteenth centuries, they are very fertile. The livelihood support from the agricultural land becomes sustainable when it helps to cope with stress and shocks and enhances the capacity of people for future (DFID, 1999). Sustainability of agricultural land can reduce the vulnerability of people in climate change situations. In a post-flood situation, the river side land becomes more saline water affected because of land slope towards embankments and stagnation of saline water near the embankment for a length of time. So the land becomes useless for a couple of years. It is true that the nature creates different types of possibilities to utilize resources. People of Sunderbans never think that flood is totally disastrous if the flood water captured in the agricultural land is not for more than one day. Although HYV crops and Rabi crops are affected, the traditional paddy (*mota dhan*) grows well.

When the saline water is stagnant for more than three days in the agricultural land, there is a very little scope for preparation for the seed bed for paddy. Villagers then leave the land fallow till the next monsoon. During monsoons, farmers start draining (*dhoani*) the saline layer from the agricultural land. Though the interior villages are able to convert the land into agricultural land, the river side land becomes fallow for one or two years. The farmers near the sea and river converted this land into temporary fishery with mixed water resources, i.e., sea water and rain water. The flood water has different species of fishes which grow quickly, once saline water gets mixed with rain water. Mixing of two types of water produces huge amounts of planktons which creates ideal aqua ecology. Though villagers cannot cultivate the land, they thus get natural livelihood support. Though all villagers are not habituated to this new livelihood support, changing climate and disasters have forced them to adapt to the changing environment with minimal support.

Conclusion

We cannot stop the impact of environmental degradation. We may be able to reduce the impact of disasters. As the Sunderbans region is already affected by climate change and environmental degradation, people have no option but to adapt to the natural phenomena. As they are the migrants from the other places, they cannot migrate again. So they adapt to the whole climate change situation with their understanding and experiences. The folk stories and songs are very effectively used for communication among different communities. As most of the local people are illiterate, they understand the natural and environmental phenomena through religious activities, folk tales or songs. Religious activities in the village help the community to organize and utilize natural resources and its protections. Social solidarity and sharing of the common knowledge through entertainment of religious symbols and group activity helps the physically and socio-economically vulnerable people to adapt to the climate change.

People's adaptation with the changing environment is a combination of experience and alternative utilization of natural resources. Once an ecological system is disrupted by natural phenomena, alternative system is created which gives support to the people to survive. Though the saline flood water pollutes the agricultural land, it also opens up a new resource, and the utilization of that resource that people have achieved through generations' struggle and their own experiences with the nature.

REFERENCES

Bandyopadhyay, S. (1997). Natural environmental hazards and their management: A case study of Sagar Island, India. *Singapore Journal of Tropical Geography*. Vol. 18 (1), pp. 20-45.

Banerjee, A. (1998). *Environment, Population and Human Settlements of Sundarban Delta*. New Delhi: Concept Publishing Company.

Basu, K. (2006). Ecology and Adaptation: A Study in the Sudarban Biosphere Reserve, in The *Oriental Anthropologist*. Vol. 6 (1), 2006. pp. 95-111.

Bera, G. *et al.*, (2006). Syncretism at Sundarbans: Anthropological and Linguistic Dimensions. *The Oriental Aanthropologist*. Vol. 6(1), 2006. pp. 5-25.

Blaikie, P. *et al.*, (1994). *At Risk: Natural Hazards, People's Vulnerability and Disasters*. London: Routledge.

Centre for Science and Environment (2007). Down to Earth Feature Service, *Vanishing Islands,* January 14.

Chottopadhya, K. (1995). *The Desperate Delta: Social Ecology of Sunderbans.* New Delhi: Gyan Publishing House.

Das, A.K. and Raha M.K. (1963), *The Orans of Sunderban,* Special Series No. 3. Bulletin of the Cultural Resource Institute, Tribal Welfare Department, Calcutta: Government of West Bengal.

DFID (1999). *Sustainable Livelihoods Guidance Sheet,* UK.

Gopal, B. and Chauhan, M. (2006), Biodiversity and its Conservation in the Sundarban Mangrove Ecosystem, *Aquatic Science,* Vol. 68. pp. 338-354.

Jalais, A. (2007). 'The Sundarbans: Whose World Heritage Site?' in *Conservation and Society,* Vol. 5 (3), pp. 335-342.

Kanjilal, Tushar (2000). *Who Killed the Sundarban?* Calcutta: Tagore Society for Rural Development.

Klein, R.J.T. (2007). *Inter-relationships between Adaptation and Mitigation,* Cambridge: Cambridge University Press.

Marfai, M.A. *et al.,* (2007). "The impact of tidal flooding on a coastal community in Semarang, Indonesia", *Natural Hazards.*

Priyadarshini, Subhra (2006) Vanishing Islands: Displaced Climate Casualties Underlying Truth. *The Telegraph,* October 30.

Raha M.K. (2006). The tribal situation in Sunderban, *The Oriental Anthropologist,* Vol. 6(1), pp. 26-43.

Tran, P. *et al.,* (2007). Flood risk management in central Vietnam: Challenges and Potentials, *Environmentalist.*

UN-ISDR (2005). Hyogo Declaration, *Report on World Conference on Disaster Reduction.*

Wadia, D.N. (1961). *Geology of India,* McMuller & Co., London.

Zockler, C. *et al.,* (2005). The Indian Sunderbans: An important wintering site for Siberian waders. *Wader Study Group Bulletin,*108, pp. 42-46.

15

Reasons of (Under) Development

Conceiving the Tribal Problematic

M. Amruth

In this chapter, an attempt is made to elicit how indigenous populations were imagined in the standard narratives of civilising and developmentalist discourses in the Colonial Travancore. I shall proceed to do this by analysing the official narratives in the context of the attempts made to control shifting cultivation in the State. For instance, a difference is effected in these discourses for distinguishing Hill-men from caste peasants in a pursuit of controlling shifting cultivation in the late nineteenth century, and the early decades of the twentieth century in Travancore. These constructions had implications to the manner in which they are conceived as governable subjects in the emergent discourses of progress and industry in the State. In these discourses of work and wealth, Hill men (tribal) populations are considered differently from that of caste populations. Further, within the category of 'tribal' many sub-categories were imagined, ultimately categorising the population in terms of their ability to involve in productive labour. Introduction of these categories and the presumptions behind these had consequences for the present day constructions of the 'idea of tribal'. These assumptions and the administrative measures based on these assumptions lead to policy formulations depriving most of the tribal communities.

A Separation of Man and Nature

The changed role assumed by the British East India Company

around the turn of the eighteenth century as a trading agency to that of the one exerting political control over populations and territory was a significant event in recasting the relations of human and nature in the territories. The devices of governance, which were introduced subsequent to the annexation, were aimed at directing human energies through modern enterprises and means to convert elements of nature into various artefacts for human consumption in and outside Travancore. In this process nature was reconstituted as an 'other' as is reflected in the observation made by the Conservator of Forests, T.F. Bourdillon, in 1893 in Travancore:

> . . . .the conditions of existence, owing to the high temperature and constant moisture, which encourage the growth of all kinds of vegetation, are very favourable. Man need only scratch the ground and sow paddy or plant a few mangoes, jacks and coconut trees, and for nine months in the year he can remain idle. The standard of comfort also is very low, as the wants of life so easily supplied and consequently, though there is a great deal of poverty there is no suffering (Bourdillon, 1893:12).

Here, Bourdillon accuses the excesses of Nature for inflicting undesirable influence on the humans. Nature and human nature are distinguishable, but observed as intimately related as the former has an undue influence on the latter. Inferior intellect, incapability for 'continuous and sustained labour' and lack of material comforts were considered as results of the characteristics of the human nature as shaped by native nature.

The above are not representative of momentous statements made in the history of Travancore as there is precedence to such thinking; for instance, still two decades before Bourdillon was to pen these opinions, Rama Varma, the First Prince of Travancore, made heralding remarks in 1874: "The happiness, prosperity, power and glory of a society or nation may always be measured by its well directed, well conserved and progressive industry. Natural gifts are unboundedly very important in the advancement of these, but unutilised by human industry; they are of little service (Varma 1874:1). He further observed that ". . . It is the very luxuriance of nature which has, in very deep in measure, stunted the spirit of progressive enterprise and industry in India ..." (Varma, 1874:2).

Nature's influences were considered as a hinderance to the progress of the humankind; nature was a persuasive deceiver. Nevertheless, it was man's fault to be enchanted by the excesses of nature, and it was man's inability to free himself from the deceptions of nature's contrivances. The will of the man had to be directed to appropriate the excesses of the nature with the modern times. Continued appropriation of the Nature's resource fulness necessitated a transformation in the attitude. The solution to this material depravity lay, according to these observations, not only in '*Othering* of the Nature' but in '*Othering* One's Own nature' and the need for its transformation. This conception of 'own nature' as problematic and necessitating change for generating labour and wealth was a key feature of the developmental modernity inaugurated in Travancore as elsewhere in British India under the colonial conditions (Raju, 2003).

The possible means for achieving these transformations from native natures to modern natures included successful deployment of modern knowledge/science such as forestry, agronomy, fishery and arts and crafts of converting nature into fine artefacts with the aid of technology and industrialisation. In the course of valourisation of these new forms of knowledge and the methods based on them, some of the systems of production as practised by natives and its practitioners alike came to be looked upon as 'Primitive'. These modalities of incorporation of territories and population into governable and wealth producing assets of the State are not always successful in case of the populations of the hilly tracts. I would further argue here that the logic of separation of human and nature achieved in this discourse functioned as reasons of government by restricting and altering the livelihoods of indigenous people and depriving their right over the land. In the forthcoming section, I shall illustrate how new legal categories and idioms were invented to appropriate the potentials of these indigenous populations and to contain their activities in the forests.

Cultivating the Modern and Dissuading the Primitive

Let us examine the case of regulation of the forest adapted cultivation popularly known as shifting cultivation–to contrast the way the State treated the hillmen from that of the caste groups. The regulation of shifting cultivation is chosen as an example mainly because the differential treatment is glaringly evident in it. Shifting cultivation was one of the many 'native practices' that met with fierce objection from both foresters and revenue administrators during the colonial rule throughout British India on a variety of grounds (Pouchepadass, 1995). The cultural practices of shifting cultivation in the subcontinent varied across regions in terms of the assortments of crops raised. However, the features that distinguished them from other forms of cultivations were the seasonal shifting of sites of cultivation and alternation in the swidden and fallow periods. Depending on agro-climatologic features of the region and availability of land for cultivation, the length of the fallow period varied between 3 to 12 years. The reason for categorising the practice as something objectionable is to be analysed as a colonial process involving normalisation of forest and creation of micro-geographies for which, reordering of labour, land and production were necessary.

Hill cultivation was referred to as *cherikkal* or *malamkrishi* in the official accounts of Travancore.[1] Bourdillon had noted that the shifting cultivation was practised by both 'Hillmen and Low-country people' (Bourdillon, 1893:117). This distinction of hill cultivators into two was the basis on which government differentially regulated them. In the case of *cherikkal* cultivation, up to three crops were sequentially raised. The first crop was usually paddy; the second either paddy or other cereals or sugarcane. The third crop was cereals (other than paddy), ginger, yams and plantains (Aiyer, 1913). During 1890s, the annual extent of land under hill cultivation was estimated to be approximately 50,000 acres. Around the 1870s, it is estimated that this would have been 'not more than 10,000 acres a year' (Bourdillon, 1893:160). It was estimated that the maximum extent of hill-cultivation was in the hilly tracts of Thodupuzha taluk[2] and in Ranni river basin. A total of 50,000 acres was annually cleared for hill cultivation. The fallow period was 10 years, the total area devoted for hill cultivation amounted to 500,000 acres or 781 square miles which is roughly the total area of secondary forests in Travancore.[3] According to Bourdillon, the secondary forests, which have a low timber value, was a result of hill cultivation.

Language of Persuasion: Reasons in Disguise

The system of hill cultivation was wasteful because it "... has destroyed immense areas of valuable forestland in the past, while its, further, continuance in the present to prevent the growth of good timber, Bourdillon observed in his report (Bourdillon, 1893:115). The arguments against hill-cultivation were founded on amazingly diverse reasons ranging from loss of soil fertility, low productivity, encouraging idleness and increasing morbidity among natives, to inefficient expenditure of time and labour. The conservator eloquently spelled out the grounds of his objections to the system of hill cultivation explicitly as below:

> This system of hill-cultivation is most wasteful, and has destroyed immense areas of valuable forest land in the past, while its further continuance in the present prevents the growth of good timber. The land selected for first instance of clearing and cultivation has always in Travancore been in such as is covered by heavy moist forest .... Thus the area of heavy moist forest has rapidly diminished, its place being taken by dense thickets of reeds, or by tangled masses of thorny creepers through which the plants of valuable trees have great difficulty in struggling. Where once there grew a forest of immense trees more than 100 feet in height with a rich undergrowth of younger trees, there is now a wide expanse of lantana, reeds, thorns and long grass. Hill cultivation is wasteful, therefore, because, in the first instance, large trees are felled and burned to make way for the clearings (Bourdillon, 1893:115).

Obviously, the prime objection to shifting cultivation was on the grounds of damages inflicted upon the timber growth; which is indicative of the utilitarian calculus. Another objection was the perceived negative impact it might have on the extension of commercial plantations, i.e., shifting cultivation had to be discouraged so that "more land would be available for coffee cultivation than would be in few years if the burning of forests [was] permitted to go on unchecked".[4]

At this juncture, it would be rewarding to look at how a planter acted in forests. "...in Peermade the planter's first job was to clear a small area of undergrowth, probably by a stream, where he could erect a few huts for his labour and himself, surrounding the enclosure with an elephant trench. Then, with one or two dozen coolies he would drive lanes through the forest to mark limits to his clearing within this area the timber was felled and burned, or left to rot..." (Lovatt, 1972:9-10).[5] Then apparently the process of hill cultivation is not very different from planting as both involved clearing of forests. Let us look for the finer details which made this mode of cultivation objectionable:

> ...the clearing of the land, and its exposure to sun and rain entirely changes the character of the soil.... Even if the land is never sown but is allowed to revert into jungle at once, the whole character of the vegetation is changed, and in the place of the original magnificent forest growth are found fast growing soft wooded trees of no use whatever. Generations, nay, centuries must pass away before the ground is again capable of supporting its former wealth of large trees (Bourdillon, 1893:115).

Although for a casual reader, these arguments might appear as if they are hitched on the colonial desiccationist discourse, the author lays bare for us, his scales of measurement by invoking the phraseology such as 'fast growing soft wooded trees of no use' and 'wealth of large trees'—a clear reference to the decreased timber value of the forests.

Plantation by forest department and private parties and extraction of timber required clearing of forests as in the case of shifting cultivation; then why the former was favoured over the latter? One of the clues available to us in the above quotes is that the former was capital-oriented productive activity, catering to the overseas markets, whereas the latter was a local activity considered as 'inefficient' and 'wasteful'. While the former catered to the market demand, the latter was mostly for self-consumption. Another clue is that the former came to represent 'skill' and 'enterprise' while both these were wanting in the latter. Yet another reason was that forestry and planting involved systematised scientific method while the shifting cultivation was done by unskilled 'natives'. Not only was the timber value of the forests affected but also the value it generated in its place was considered as meagre. And "...this system of hill cultivation is wasteful because, under it, the same area of land will only support about one-fourth of the population which regular cultivation would feed" (Bourdillon, 1893:116).

It was estimated that "in permitting heavy forest to be felled for cultivation loses about Rs. 200 in their share of the value of the timber destroyed, in exchange for 1 rupee realised by the tax collector, over every acre of heavy forest land thus cleared" (*ibid.*, 116). Besides the loss of revenue, there are also cases of the cultivators evading payment of revenue or rent (*ibid.*, 116). Rent was calculated by comparing the loss incurred due to hill cultivation in terms of the timber destroyed. Here the assumption is that timber and the agricultural crop raised in hill cultivation are similar. It was conveniently forgotten that while the hill cultivation is an annual production activity, which reproduces native labour, the latter's gestation period spans to generations. The conservator here attempted to substantiate that the productivity of shifting cultivation is much lesser than that of the settled cultivation. Let us leave these expressions of economic rationality of the Eurocentric kind and calculations of legitimisations, and proceed to focus on how they portrayed their 'Other/Idle Natives'.

Portrayal of 'The Other'

One of the most consistently articulated arguments against the practice of shifting cultivation was that the practice with its lack of sophistication encourages 'idleness' of the 'natives.'

> ...this system is objectionable because it encourage the people to be idle. They go out into the forest in the dry weather and make their clearings, and after a time they return and reap the crops, watching them for a short time before they ripen, and all together they spend 3 months over the work, but for the other 9 months they have nothing to do, and these 3 months of work keep the cultivators and their families in food for a year (*ibid.*,115).

Incompatibility between this system of crop production, and

plantation/settled cultivation is well engraved in the above quote. This incompatibility is constructed by anchoring it on the desire for maximising productivity and rationale of governance. It was not only the forester who perceived shifting cultivation as promoting idleness, missionaries also felt so. Rev. Samuel Mateer maintained that, though the crop was excellent and crops cultivated were similar to that of the plains, the factors such as loss of timber, and the laziness it promotes alone made it objectionable:

> This mode of cultivation yields a larger return for the same amount of labour than permanent plough husbandry, but is highly destructive of valuable forest lands. Their migrant habits arise partly from laziness : it is easier to cut down and burn new forest than to rear cattle, plant trees, manure land, and build houses. They, therefore, prefer this savage life, but should be encouraged settle if possible: only by such means can they be reclaimed to civilisation and education, as has been done farther north (Mateer, 1883: 64-65).

Sources of inefficiency are found to be rooted not only in the mode of shifting cultivation alone but also in its context. This prompted the conclusion that 'everything is against the system of Hill-cultivation, and it has nothing to recommend it'.

The manner in which the Nature was perceived to be influencing native nature was diverse. Sometimes, the reason was pitched on the ill-health of the labourers as if there is a 'care' flowing to them. These influences can be in the form of susceptibility to diseases caused by unhealthy elements of Nature or to attacks by *ferae naturae.*

> . . .Hill-cultivation is bad because those who follow it are exposed to the risks of fever or the attacks of wild animals, and the country is kept unhealthy to an extent that would not occur if the land were regularly cultivated. With easy work and abundant supply of food, it might be supposed that all the conditions necessary for a rapid increase of population were present, but as a matter of fact the population increases at a very slow ratio. To get to their clearings the people have to pass through the rank scrub-

> jungle reeking with malaria; they have to sleep out in the open without proper shelter: sometimes the wild elephants which come to feed on the crops kill one or two: sometimes man-eating tiger appears, and ceases a victim here and another there. At one time, when the crop is just gathered, they live in great abundance, at another they were starving for weeks together while the grain is ripening, and then they have to subsist on the seeds of the *eentha* (*Cycas circinalis*), the pith of the bastard sago (Caryota urens), or such roots as can be found in the jungle, while, should the crops fail, many die from insufficient food. Thus the average length of life of the Hill-cultivator is short (Bourdillon, 1893: 116).

There is a convergence of the views on shifting cultivation expressed by forester, missionary and the Prince/State. Besides, it is well-known that the colonial anthropology/ethnography has also carved out an image of leisure-mongering and lazy native. One is tempted to propose that such portrayals did not confine to the forestry alone but it was also part of a general developmentalist discourse. Shrouded in these arguments are the latent elements of modern economic rationale in which colonisers were also entrenched in. But, what is still interesting is to look at how this agenda was concealed by bundling them up with other arguments and concerns for achieving legitimacy and currency. That is, the arguments are bundled together as more scientific, legitimate, idealised and acceptable categories of modernity, which had ready buyers. Prevention of shifting cultivation was not so much born out of scientific forestry but out of Orientalist discourse enabling resource appropriation and wealth transfer to Britain. Some of the 'reasons' identified in this section clearly reflect the nature of colonial conception of labour, efficiency, nature, and native nature. Having listened to varied reasons devised to banning shifting cultivation, let us now examine how the shifting cultivation was controlled.

Regulations and Control as Strategies of claiming the Territories

The directives and proclamations restricting shifting cultivation are permeated with the idea of claiming the territories and bringing these territories to the logics of control and governance. Let us examine here these reasons of government closely. One of the early attempts of controlling shifting cultivation was in 1866 when its prohibition was proclaimed.[6] A further regulation was issued in 1870 laying down rules for hill cultivation, according to which people were allowed to cultivate without permission the grasslands, reed forests and marshy places belonging to Government, but they were prohibited from clearing land on the "hills where clouds rest, or at the jungles at the lower elevations where there are Sircar trees such as teak, black wood, etc." Further, people were banned from clearing land "where there are trees of more than 10 virals quarter girth".[7] The rules thus laid down also necessitated acquiring permission from revenue and forest officers by undergoing a rather lengthy procedure, and the permission granted was not to exceed three years.[8] However, location specific measures to control shifting cultivation were also taken as is evident from a regulation, which envisaged controlling cultivation at Thovala in 1876.[9] According to this regulation, cultivation in the hill tracts of Thovala was restricted through licensing. The reason for such restrictions was the concern on the clearance of the hill vegetation which negatively affected the flow of the rivers that irrigated the agriculture in the plains.

Stipulations for the protection of royal timbers, and distribution of permits by the Conservator, were also necessary in the case of cultivation on hill tracts or *cherikkal* belonging to the landlords (*jenmis*).[10] Though the assessment of tax on *cherikkal* (called *malavaram* or *vilameladi*) was made mandatory, to be paid as a share of yield, it was difficult to enforce. This was because the *cherikkal* were located in higher altitudes prevalent with malaria and this dissuaded the officers from visiting these hill slopes for collection of tax. Moreover, the 'system of assessment was. . . . open to serious abuse at the hands of low paid village revenue officials. Either large portions of cultivated lands wholly escaped assessment, or a portion of the tax collected on them did not find its way into the Public Exchequer.'[11] As early as in 1835, a regulation was issued to foreclose the possibility of tax evasion by the cultivator. Prescriptions were also issued to enhance efficient collection of *malavaram* taxes.[12]

The forest regulation of 1888, rules and subsequent amendments had provisions discouraging clearing of forest land for shifting cultivation.[13] The previous proclamations of the land revenue assessment made *cherikkal* to be assessed in terms of the ordinary *puduval* rules, where assessments were high.[14] On the basis of a representation made by cultivators, a provision for reassessment of the *cherikkal* lands was made in 1905. In 1907, a further regulation was issued for final settlement of *cherikkal* accompanied by augmenting the revenue assessment and making the registration of the *cherikkal* land mandatory.[15] Following the completion of the *cherikkal* settlement in 1911, an order was finally issued for prohibiting clearing of any new land for hill cultivation, which described the *cherikkal* cultivation as 'fugitive cultivation' capturing most of the images of shifting cultivation cherished and propagated by the State.[16]

Instances of these regulations provide us an opportunity to examine the implicit and explicit reasons quoted for controlling the practice. For instance, the hill cultivation was discouraged to favour the coffee cultivation as it is evident in the proclamation of 1866. It is also worth noting that in the early decades, the objections against hill cultivation were mostly grounded on the difficulty in collecting tax, not so much on the loss of timber. The arguments of timber loss and threat to conservancy were distinctly a development of the later phase. It was in one of the proclamations passed in 1857 that the damages to the forests and timber were explicitly pointed out as a reason for regulating the shifting cultivation.[17] The regulation of 1870 and the forest regulation of 1888 were informed by the influence of the forests on local climate besides the economic rationales of the timber protection. That is, from short-run rationale to long-run rationale; from immediate gains to gains in the future. Over the period, apparently, the regulations have become increasingly specific, punitive and stringent; following the regulation of 1911, this tendency culminated in a total ban on any further extension of the land under *cherikkal*.[18] As a result of these measures, the extent of *cherikkal* lands were constantly declining; land under hill cultivation has declined from approximately 10,000 sq. miles at the beginning of the nineteenth century to 790 sq. miles towards the last decades of the century. This reduction was partially due to the sale of the government *cherikkal* to establish commercial plantations by private entrepreneurs. There were peasant oppositions, often leading to violent conflicts against transfer of *cherikkal* lands to planters, in the first decades of twentieth century.[19] Most of these incidents were followed by spread of rubber cultivation in the lower slopes of the *Ghats*.

If the subtropical plantations of coffee and tea affected the tribals who practiced shifting cultivation, it was the marginal farmers and agriculturists who were affected by the enclosure of lands in the lower altitude. Around 1905, almost five years before the final notification of ban on the hill cultivation, when Bourdillon was still the conservator, government sought expert opinion on granting lands in the lower elevations for cultivation of rubber. Bourdillon, while replying to the query stated that, ever since his working with the department he has been making efforts to consolidate forest territory by defining the government owned forest area to "eject the squatters and cultivators who may be found to have no title to the lands. The work of settlement of forest reserves and the suppression of temporary cultivation, however, has taken more time than. . . . [was] anticipated".[20] He was optimistic as the process was approaching completion and it was "expected that in about 3 years hence, all the lands suitable for cultivation will have been selected and finally settled and temporary cultivation stopped altogether."[21] While referring to the requests for forest land for rubber cultivation, Bourdillon has resorted to quote extensively from the forest policy of 1894 in British India. Where he explicitly suggested, following the British Indian forest policy, that 'honey combing' of the forest lands for cultivation has to be discouraged at all costs and temporary or shifting cultivation needs to be curbed completely. In doing so he also admitted the possibility of dereserving forest lands such as 'minor forests' and 'pasture lands' for the purpose of settled cultivation. He also maintained that in extreme cases, the other types of forests also may be dereserved considering the specificity of the cases. Subsequently the government adopted these recommendations.[22]

What is interesting is that the State was discouraging shifting cultivation at a time when it was all set out to promote cultivation of foodgrains by offering generous tax concessions and financial benefits.[23] Travancore was an importer of foodgrains throughout the period. Augmentation of food production was a major concern of the State. In the early decades of the nineteenth century, precisely from 1817 onwards, there were attempts by the State to provide tax exemptions for encouraging people to settle in the hill tracts that were abandoned due to the depredations of wild animals.[24] In subsequent years, similar proclamations were made with a promise of more generous concessions for opening up forest for cultivation.[25] However, these measures failed to produce expected results due to crop destruction by the wild elephants.[26] Generous land grants and financial aid for reclaiming lands from backwaters were instituted for increasing cultivable land from last quarter of nineteenth century itself.

A Rule of Difference

Shifting cultivators were generally classified into two: 'hillmen' and the 'low country people'. One of the implications of this distinction is administrative control effected through categorisation. Having said this, it requires further examination to understand the functionality of these categories in governance. It had been observed that the "low-country people. . . . could find other means of subsistence if they liked" and that these agents need to be discouraged from continuing the practice.[27] It also implied that low-country people were to participate in the production process of the State in a different way from that of the 'hill men'. Bourdillon furthered this argument supported by the following observations:

> The former [Hillmen] confine themselves chiefly to the hills and forests of the interior, and clear land which is often valueless to us, because it lies in position as to be inaccessible for the removal of its timber, or because it contains trees of but little worth. Further, there is a tendency among these people to settle down in the same places, and not to move about so much as formerly. This is a tendency which is by all means to be encouraged, and as the Hillmen are very good axmen, and are well acquainted with every track and by-path in the forests, their services can often be utilised by us. Besides this they, as a rule, only raise enough food for their own requirements, and their number do not exceed 8,000, so that the damage done by them is small in comparison with that effected by the low-country people. It is quite

> impossible to entirely prevent these Hillmen clearing land in the forests, and the only way to deal with them is to induce them to make their clearings in the less valuable land, where the timber is of no worth, and its distance from the low-country would render the cost of working down the timber prohibitive, and to induce them to concentrate their operations in one place by allotting land to them. I do not consider it advisable [sic] to endeavour to tax or coerce them. (Bourdillon, 1893:117-118)

State differentially perceived and distinguished between tribes and low country people. Tribes were considered as thoroughly acclimatised, sturdy and intuitive people capable of surviving in the wilderness, whereas the caste groups of peasants were considered as lame and susceptible, requiring State patronage for walking their way to civilisation. In the approach to hillmen, it is the restrictions of their activities and utilisation of their skills that were more important. Hillmen were the objects to be policed, whereas the low country people were to be patronised and provided. At the same time low country people were to be made into disciplined and docile bodies practicing settled cultivation; as for tribals, they were to be let loose in the forests except pressurising them to abandon shifting cultivation and causing damages to valuable timber. State also expected sizable number of them to live as rugged labourers.[28]

The administrator's convictions about the shifting cultivation cannot be mistaken. To them it was a practice that was to be discouraged. Irrespective of the differences in the categories of people, all were expected to become settled cultivators. The 'low country' caste communities were expected to settle down, while the needs of the hillmen were viewed more sympathetically and they were to follow their low country counterparts on the long run.

Rules for the treatment and management of hillmen in Government Forests under the forest regulation and its subsequent amendments in 1911 have restricted their cultivation to the area earmarked by the Forest Department. This allowance was made on the condition that they agree to work for the department at reduced rates of wages. This was also a time when Forest Department was in need of more labour force; for, it had been engaging in operations in interior forests. The primitive knowledge possessed by the hillmen was useful for guiding the forester for deploying his modern knowledge for normalising the nature. The need for able and skilled labour (axe man) and tracker in the forest were the implicit reasons for recommending the continuation of hill cultivation by hillmen. Moreover, their population was smaller and they cultivated in the forests of lesser timber value. This strategy had encouraged them to settle and practice hill cultivation in areas of low effective timber value. Clearly, the logic of regulation was the economy. But, the arguments were clubbed with ideas such as civilizing, progress and improvement.

The constant annoyance of the administration towards hill cultivation was due to the 'primitive technology' involved in the shifting cultivation and its incompatibility with the system of taxation or revenue administration. It is primitive because it was incompatible with the forestry ideals of timber conservancy. What was imagined was a move away from the idea of 'shift' to 'sedentary'; they were the signs of Eurocentric progress and scientificity. A similar idea of progress and evolution was imagined by colonial ethnographers, and natives and tribes were placed at lower scales of progress. To the colonial government, controlling shifting cultivation had various meanings; curbing the destruction of forest and protecting the valuable timber, 'civilizing' the primitive who lived in subhuman circumstances, promoting the settled cultivation and enforcing a land taxation system that expected every piece of land to have a legitimate and recognisable owner/tax payer. Control of *cherikkal/malavaram* also resulted in settling of the nomadic communities and drawing them to monetised economy. Moreover, there are evidences which suggest that the British benefited from the additional labour force made available due to prohibition of shifting cultivation.

The skills and familiarity of the hillmen in the hilly terrains were lauded by the colonial writers in different capacities. Muthuvan, Kanikkar and Mannans were considered as very good trackers. Without the services of Urali population, wild elephants could not be captured and tamed. Without hillmen valuable forest produce could not be collected and taken to the markets. Although compulsory labour called *Uliyam* were abolished way back in the 1860s, hillmen were to work for Forest Department at reduced wages as per the notification for management of hillmen proclaimed in 1911.[29]

As observed in the previous sections, the portrayals of the hillmen were limited at producing an impression of State patronage for making them more useful and submissive but not so much to effect progress among them. As a result, these tribes did not get objectified in the development discourse. They were objectified only as rugged trackers and axe men or manual labourers; and therefore, there was little scope for them becoming the subjects of developmentalist discourse. In fact, what they wanted was only few concessions as enjoyed by the lowest caste at their social proximity. Let me elucidate the point made here.

Though forest department had powers over the tribes, civilising was not its mission. From what follows, we can rightly discern that the interest of the Department was in tribal people's ability to labour; as they were useful in appropriation of resources.[30] In 1909, the prominent ethnographer of 'Cochin Castes and Tribes' noted that:

> .... but for the Kadars and the elephants, the Cochin forests would have been useless. Both are the denizens of the hills and forests, and when properly utilised, they are invaluable help to those who wish to extract the priceless treasures of the forests. But for the elephants none could lift a log, and if it were not for the Kadars, none dare move an inch in the forests. During the forest working season these men are employed by the forest contractors for collecting minor forest produce (Iyer, 1909:15).

This explains the State's perception about the indigenous population. *Kadars* were one of the 'hillmen' groups residing in the forests of Anamalais and a relatively scarce population in Travancore territory. There are sufficient indications to believe that the tribes were subjected to compulsory labour, without payment, in the forest operations, cardamom gathering and cultivation. In 1866, forced employment of the tribes in the watch stations without pay was abolished. It could be assumed that some privileges were provided for those who were already engaged in such works. It was stated that "[a]t present the only remnant of this modern *socage*.... existing under either the Forest or Cardamom

Departments, is that the Hillmen are expected to work for us at a reduced rate, in consideration of their being allowed to clear land in the forests for their crops."[31]

To the Forest Department, tribes were useful objects; as trackers and as labourers they provided the much required assistance in the forestry operations in the remote forest areas and they collected most of the 'Minor Forest Produces'. Without tribes, forest would not have been a 'resource'; rather, it would have been a wasteland. While charting out the plan for future management of the forests in 1893, the Conservator of Forests remarked on the utility of the hillmen:

> These people do comparatively little harm now, in fact, I may say, none, unless they fell virgin forest. . . . they are very useful in building huts, cutting paths, collecting wax and cardamoms, assisting in the capture of elephants, tracking and acting as guides. They should, therefore, be allowed to remain where they are (Bourdillon, 1893:209).

What is interesting for us here is not that the author contradicted his own opinions about the hillmen expressed elsewhere in his report, but that all of a sudden hillmen assumed some virtuousness; they become a welcome presence in the forest; clearly, owing to the possibility of their incorporation in the material processes. The subsistence-level shifting cultivation practiced by hillmen had been constantly opposed by the government from the 1860s and more elaborate and stringent measures for controlling, sedenterising and disciplining the hillmen were envisaged in forest regulation. Bourdillon's observations such as the one quoted below were soon translated into rules:

> I am told that on the Annamalais the Kadars have been allowed to remain in the forests on the condition that they fell no new forest, but confine themselves to those parts which had been cleared by them before. This arrangement has been very successful and these hillmen are contended and prove themselves of great service to the forest-officers. Some such restriction should be introduced here under

heavy penalties of fine or even expulsion for its infringement (Bourdillon, 1893:209-210).

In the title of the regulation, 'Rules for Treatment and Management of the Hillmen in Government Forests and Reserves', hillmen were equated with that of non-human forestry objects by invoking forestry terminologies such as 'treatment' and 'management'.

The recommendations of the Conservator and the rules enabled exercise of modern disciplinary power on the tribal population. Some of the processes in extending the disciplinary power included: enumeration, sedenterisation and colonisation. Their continued stay in the forest was made conditional to their supply of manual labour in the forestry work.[32] The enumeration, registry and surveillance of them were effected through galvanising the headman into the economic and institutional process of the Forest Department. Let me quote a relevant portion of the rule on the enumeration process: "Hillmen residing in Government Forests and Reserves shall have their numbers registered in the office of each Forest Division, and for this purpose the headman of each settlement shall report to the range officer concerned, once a year or whenever called on to do so. . . ."[33] Continuation of the headman depended on the discretions of the Divisional Forest Officer; the rule states that ". . . when headman fails to carry out any of the duties, . . . the Divisional Forest Officer may call upon the adult male members to dispose him and elect another competent man to take his place", and all the decisions taken by him had to be ratified by forest officials.

Traditional livelihood activities such as shifting cultivation, hunting and gathering, collection of fuel wood, and small timber, have became a 'generously offered concession' and 'privilege' by the government. Further, in order to be eligible to 'enjoy these privileges' they are obliged to sell their labour to the Department 'whenever called upon to do so, at prescribed rates of wages.'[34] They were to "assist the Police, Revenue or other officers of the Government in preventing crimes or in detecting offenders who may have sought shelter or refuge in the Government Forests or Reserves."[35]

All walks of the tribal economy were either controlled by or channelled through the Forest Department. Though the rule generously endorsed proprietorship on the produce he cultivated on the allotted land (of course without a title), he/she was not allowed to sell rice and tobacco thus produced. Moreover, they were "bound to deliver to Forest Department any ivory, elephant teeth, cardamom, wax, dammar and lac collected by them, as well as honey or any other forest produce which they may be called upon to collect and deliver to the Forest Department, in return for which they will be paid at the rates fixed" by the Department.[36] Whatever meagre expenses incurred by the Department in providing petty 'concessions' and 'privileges' were, therefore, more than compensated for by capitalising the 'usefulness' of the hillmen as (non)human resources of captive labourers. Even the traces of rhetorical phraseology such as 'improvement', 'progress' and 'onward march to civilisation' which otherwise copiously found in every pages of the Report were starkly absent when it dealt with hillmen; only their workmanship, animal like instincts, familiarity with the complex terrain and endurance in inhospitable weather were celebrated. Despite these celebrations, Rules were in general harshly worded and stringent in nature, especially so when it stated "Any hillman found to wilfully violate, or act in contravention of these Rules, or of Forest Regulation and Rules shall be liable to be expelled from his settlement in a Government Forest or Reserve, temporarily or permanently . . . besides being subjected to the penalties prescribed in the Forest Laws."[37] The hamlet had come to resemble some kind of a labour camp of the proletarianised but lesser privileged and depended at the discretions of the officials.[38]

In what appeared as a short circular issued by the Conservator in 1946, a radical change in governance of hillmen is evident. Conservator, taking cognisance of the information that "Kanikkars and other hill-men are in the habit of visiting the Palace to pay their respects and represent their grievances", redirected such representations of grievances to the officers of the Forest Department. Therefore, as a consequence, visits to the Palace got prohibited on the ground that the hillmen were under the 'protection' of the Forest Department. To the Hillmen, the ruler is no more the king but the Conservator. If at all they intend to go to the King, they should ". . . not go to palace without first obtaining the sanction of Conservator of Forests."[39] Here two different forms of disciplinary relationships come to play: 'permission' and 'control'.

Lineages of (under)Development

Following the case of restricting the shifting cultivation, it is evident that a distinction emerged between the tribes and the non-tribes and this led to their differential treatments. It was thought that tribes need to be restricted, whereas the low caste cultivators at the forest fringes need to be inculcated the habit of restraint. What we find in this section addressing the questions related to the normalisation of forests for the present and sustained supply of exchange and use values is that in this process, the subject position of the tribes and forest fringe communities changed substantially. At the macro level, they were the subjects of the king or the local chieftain, but with the advent of scientific forestry practices they became the objects and resource of the State.

In forestry, as practiced in Travancore, the foresters wielded absolute control over the tribal populations. Foresters, based on their self-identification as the bearers of a superior technology of harnessing the nature/resource often portrayed all the indigenous enterprises that failed to fit in their temporality as primitive and backward. Not only did such communities appear backward, but their 'ignorance' is also interpreted as stemming from a lack of concern for the State and master, ultimately judging such an attitude as lack of morality. Guiding the primitive forward to the new temporality of the utilitarian, futuristic scheme of thinking and morality as per the principles of the cameral forestry was a challenge that foresters of Travancore had taken up like their counterparts in British India and continental Europe.

The idea of progress invokes a notion that the future of the native society will have to achieve the standards of industrialised societies of the West. The present of the native is identified with the past of the Western society. Therefore, primitive nature of the 'native nature' is considered as a demanding site of intervention, requiring control and reformation. Here nature is a referent which is indefinitely redefined, shifted and changed on each encounter with the modern. Native men are considered as having some sort of mastery over nature but since both are primitive, the need was to create a modern nature, that is accessible, predictable and controllable; so a modern master and mastery were the needs of that hour.

As we have already seen the modes of incorporation of the indigenous population in hill tracts to the system of State controlled production was different from the modes of incorporation of the caste communities in the plains. Though every effort was made for sedenterising the indigenous population on a long run, as an immediate measure to control the 'primitive' cultivation, hillmen were brought under the direct control of the forest department. Where, through the rules laid out specifically for the Hillmen, not only were their activities in the forest controlled but they were also forced to labour in the forestry operations on compulsory basis at reduced wage rates. Cultivation was to be restricted to specified areas earmarked once and for all as hamlet areas. Any surplus generated was to be handed over to the State agencies. Population in reserve was to participate in surveillance against illegal activities in forests and function as informants. Threat of eviction, change of chieftainship of the hamlet and criminalisation became consequences of not complying with the measures of incorporation.

Justification for these acts was sought in the perceived uniqueness and primitive state of the indigenous populations. Above all, these portrayals of primitivism and invoking of the backwardness in the cultural evolution of the tribal population indicated the disenchantments and discontents of the state in their easy incorporation in the mechanisms of accumulation. For instance, they were the expert trackers and labourers in forests. Collection of minor forest produce, including ivory was entrusted to them. At the same time, these practices and its schemes of legitimisations placed the tribal populations in the twilight of the State schemes and citizenship. They were to be under the patronage of the State and were unqualified to own land of their own unlike their counterpart caste people who also practiced shifting cultivation. Thus the agrarian traditions of most of the indigenous communities were disrupted; they were incorporated as underpaid compulsory labour in the project of colonial forestry. They also were denied of the right to own and exchange land in the pretext of having not acquired necessary civilisational maturity. It is not difficult to see this logic lurking behind the current state portrayals of tribal communities in the pretext of land question.

NOTES

1. *Cherikkal* or shifting cultivation was also known as *Malamkrishi* (meaning cultivation in the hills) and *Malavaram* (referring to the taxable shifting cultivation). There were mainly two kinds of *cherikkal* in Travancore; the Sirkar or Government *cherikkal* and private *cherikkal*. In the former, the land belonged to the Government and in the latter the land was owned by private individuals. In this section, we focus on the legal or illegal *cherikkal* or shifting cultivation practiced in the government lands.
2. Hill-cultivation was absent in the coastal taluks. Out of the 32 taluks in the Travancore during this period, in 9-10 taluks had significant area under hill-cultivation.
3. Bourdillon has classified the forests of Travancore into following four classes: moist forest, secondary forest, grass forest and useless land.
4. Extract from Dewan's Mimeo on Forests,1868, in section book of revenue letters from 19 March to 22 August, 1868, quoted in Pandian (1990: 79).
5. This account seems to have been adapted from the accounts of F.G. Richardson (known as 'JAR') who ventured out in the east of Peermade for extending plantations to Vandiperiyar. See also Muthaiah (1993: 54).
6. It stated that 'the felling of forest by Hillmen or others for cultivation, if suitable for coffee plantations' (Bourdillon, 1893: 161) was forbidden.
7. Bourdillon (1893: 132), and Proclamation No. 5674 dated 25th Meenom 1045 M.E. (*TLRM*, Vol. V, 1916: 643). In proclamation No. 4167 dated 6th Meenom 1056 (M.E. 1881) a clarification was made on the conditions in the earlier notification.
8. The process of obtaining permission begins with the *ryot* making application to *provethikar* and his report on the examination of the land communicated to the Forest department through Amindar. The decision of the Conservator of forests (mentioned as 'Mr. Vernede, Conservator' in the regulation) on this matter is thus communicated to the Tahsildar who if application is eligible for permission would issue a temporary title, with specification of limits and area of cultivation.
9. Notification No. 3108 dated 1st Kumbhom, 1051 M.E. (*TLRM*, Vol. V, 1916: 654-655).

10. *TLRM,* Vol. IV (1915: 401).
11. Aiyer (1913: 122), see similar remarks in the Notification No. 6909 dated 28th Medom 1063 M.E. (A.D. 1888) published in Travancore Government Gazette reproduced in The Travancore Land Revenue Manual (*TLRM,* Vol. V, 1916: 895-96), Proceedings of the Government of His Highness the Maharaja of Travancore, and also in The Travancore Land Revenue Manual (*TLRM,* Vol. II, 1916: 817-819).
12. Proclamation dated 17th Karkadagom 1010 M.E. (Huzur Central Vernacular Records) *TLRM,* Vol. 5: 411-13.
13. It was mentioned in the regulation II of 1068 (i.e., Forest regulation of 1893) that "No fresh clearings for cultivation or for any other purpose shall be made" on the land notified as intended to be reserved. See *The Travancore Forest Manual* (*TFM* 1917: 8).
14. A series of proclamations passed from 1884 to 1888 changed the basis of revenue assessment in the State from 'produce of the land' to the 'extent of land in possession' irrespective of the fact whether it was cultivated or not. The tax payment got monetised. [For instance, see the Notification announcing the modalities for encouraging ginger cultivation in the *Malamcherikkal* cultivation and fixing the tax in money, Huzur Sadhanom No. 1894 dated 26th Kanni 1060 M.E./A.D.1885 (Huzur Central Vernacular Records). (*TLRM,* Vol. V, 1916: 790-91)] The significance of this shift in policy was that it was intended to make the ryots more enterprising by offering incentives for producing maximum from the land under possession. Considering the same rationale, it may be discerned that the regulations on the hill cultivation were part of State's desperate attempt to wield control over the elusive hill peasant who remained outside the existent system of surveillance and governance.
15. See the Order No. 849/R. dated Trivandrum, 13th February 1907/ 2nd Makarom 1082 (*TLRM,* Vol. V, 1915: 379-383).
16. Government Order No. 3427/L.R., dated Trivandrum, 16th April 1911/3rd Medom 1086 (*TLRM,* Vol. II, 1915: 818-819).
17. See proclamation dated 2nd Kumbhom 1032 M.E. (Huzur Central Vernacular Records), (*TLRM,* Vol. V, 1916: 539-40).
18. Pandian (1990: 124-45).
19. Rammohan (1996: 69).
20. Proceedings of the Government of His Highness the Maha Rajah of Travancore, The relevant portions of the reports of the Conservator quoted in the Order No. 758/R.283, dated Trivandrum, 14th January 1906/1st Makarom 1081 (*TLRM,* Vol. II, 1915: 256-560).
21. TLRM, Vol. II (1915: 256-560).

22. Order No. 4692/R. 1874, dated Trivandrum, 21 March, 1906/8th Meenom 1081. And rules based on the recommendations of the Conservator in this case have been issued within three months. See order No. 4696/R.1874 dated 21st March, 1906.
23. For instance, see the gazette notifications offering substantial loans for the reclamation of backwaters for paddy cultivation; Notification No. R505 dated 8th, 1063 M.E, Travancore Government Gazette, (*TLRM*, Vol. V, 1916: 912) and the government sponsored schemes for promoting paddy see Notification No. 1063, dated 13th Mithunam 1063 M.E., Travancore Government Gazette (*TLRM*, Vol. V, 1916: 906).
24. A near total disarmament prevailed in State during the these early decades of nineteenth century, following the revolt of 1808 said to have caused a decline in agricultural production, invasion of natural vegetation and wild animals forcing people to retract from forest fringing settlements towards plains (Ward and Conner, 1863: 43).
25. The concessions thus promised were generous with provisions such as exemption from the assessment for 10 years in case of cultivation of unregistered lands (*puduvals*) and an exemption of the cost of improvement during the registry of land after 10 years. In case of old registries, the tax exemption was for nearly 6 years for recultivation. Land cultivated with annual crops such as plantains and tuber crops were exempted from paying assessment and ryots were also promised money loan on security (*TLRM*, Vol. IV, 1916: 250-253)
26. *TLRM*, Vol. IV (1915: 273)
27. Bourdillon (1893:117)
28. Treatment of tribes and the plains men were different to the Colonizer. In case of plain's men they were considered as feminine/ dealable while the tribes were portrayed as masculine/boyish/ usable. See also Skaria (1998: 214-215).
29. Mateer (1883: 78-79), a missionary with London Mission Society described the oppressions and fleecing to which hillmen especially Mala Arayas were subjected to by the local chieftains and officers of various departments for extraction of their free services for various purposes.
30. The improvement and progressing discourse also did exclude and look for utility by dispossessing.
31. Bourdillon (1893: 161), emphasis added.
32. See rules (*TFM* 1917: 132-136).
33. See rules (*TFM* 1917: 133).
34. The wages, however, were reduced wages as evidenced from the

following statement: "... modern socage, or "*ûliyam*" as it is called, existing under either Forest or Cardamom Departments, is that the Hill-men are expected to work for us at a reduced rate, in consideration of them being allowed to clear land in the forests for their crops" (Bourdillon 1893: 161).

35. Section 16 of the Revised Rules for the Treatment and Management of Hillmen in Government Forests and Reserves (*TFM* 1917: 132-136).
36. Section 18 of the Revised Rules for the Treatment and Management of Hillmen in Government Forests and Reserves (*TFM* 1917: 132-136).
37. Section 19 of the Revised Rules for the Treatment and Management of Hillmen in Government Forests and Reserves. (*TFM* 1917: 132-136).
38. See Amruth and Gurukkal (2007: 131-132).
39. Conservator's Circular No. 3 of 1121, dated 8-10-1945.

REFERENCES

Aiyer, P. (1913). *Travancore 1883-1911 Final Report: Revenue Settlement*. Trivandrum.

Amruth, M., and Gurukkal, R. (2007). Spatiality of Subsistence and the Human Ecology of the Landscape: Towards Self-RegulatoryForest Communities. In Ramesh, B.R. & Gurukkal, Rajan (Eds.), *Forest Landscapes of the Southern Western Ghats, India: Biodiversity, Human Ecology and Management Strategies* (pp. 97-140). Pondicherry: French Institute of Pondicherry.

Bourdillon, T.F. (1893). *Report on the Forests of Travancore*. Trivandrum: Travancore Government Press.

Iyer, L.A. Krishna (1909). *The Cochin Tribes and the Castes—Volume-I*. Madras: Government of Cochin, Higgin Bothom & Co.

Lovatt, H. (1972). *A Short History of the Peermade/Vandiperiyar District*.

Mateer, Samuel (1883). *Native Life in Travancore*. London: W.H. Allen & Co.

Muthiah, S. (1993). *A Planting Century: The First Hundred Years of the United Planters' Association of Southern India, 1893-1993*. New Delhi: Affiliated East-West Press Private Ltd.

Pandian, M.S.S. (1990). *The Political Economy of Agrarian Change: Nanchilnadu, 1880-1939*. New Delhi: Sage Publications.

Pouchepadass, J. (1995). British Attitudes towards Shifting Cultivation in Colonial South India: A Case Study of South Canara District 1800-1920. In D. Arnold & R. Guha (Eds.), *Nature, Culture, Imperialism: Essays on the Environmental History of South Asia*, (pp. 123-151). New Delhi: OxfordUniversity Press.

Raju, S. (2003). Developmental Modernity: Man and Nature in the Discourse of Wealth and Labour. *Contemporary India: Journal of the Nehru Memorial Museum and Library*, 2, 45-74.

Rammohan, K.T. (1996). *Material Processes and Developmentalism: Interpreting Economic Change in Colonial Tiruvitamkur, 1800 to 1945*, (unpublished Ph.D Thesis) University of Kerala.

Skaria, A. (1998). Being Jangli: The Politics of Wilderness, *Studies in History*, 14, 193-215.

TFM (Travancore Forest Manual) (1917). Government of Travancore. Trivandrum.

TLRM (The Travancore Land Revenue Manual), in five volumes. Vol. II, 1915, pp. 256-260.

Varma, R. (1874). *Our Industrial Status*, Trivandrum Debating Society, CMS Press. Cottayam.

Ward, B.S., and Conner, P.E. (1863) [1994]. *Geographical and Statistical Memoir of the Survey of the Travancore and Cochin States Executed under the Superintendence of Lieutenants Ward and Conner from July 1816 to the end of the Year 1820. Vol. I.* Reprinted State Editor, Gazetteers Department, Trivandrum.

16

The 'Political' Dimension

In the Protest Movements of the Indigenous Groups

Geetha Devi, T.V.

The emergence of tribal protests in India presents a history that has a bearing on their naturally inhabited area, i.e., forest, its depletion, loss and further displacement from it. Though a variety of reasons can be attributed to the tribal conflicts in India in the backdrop of the dwindling natural resources, it is seen that the main thrust of these movements that happened during the second half of the twentieth Century were (i) the deforestation activities that have begun during the colonial period; and (ii) the development policies of various governments in India especially after independence. This chapter aims to focus on those protest movements that have taken place in the above-mentioned contexts in India. The main intention of this chapter is to do a theoretical exposition of what is 'political' (Sheth, 1997) in these resistance movements. Following is an account of some case studies regarding the movements that took place after 1950s, which this study has taken up for analysis.

The earliest form of people's organized protest was started in the 1970s, known as *Chipko* (to hug) movements, which were held against the State governments in the Chamoli district of the Himalayan region (Guha, 1989). This kind of a situation was created in the context of the taking over of India's forests by the governments during the colonial period and after independence. The forests were under the control of the local village communities till the second half of the nineteenth century and there was little state intervention.[1] The villagers of Mandal resorted to the kind of 'Chipko' protest, when their request for using the forests for local needs was denied by the Uttar Pradesh Government in 1973. The *Dashauli Gram Swarajya Sangh* (DGSS), a co-operative organization based in Chamoli district, approached the government for getting allotment of ash trees in order to make agricultural implements, which was refused by the forest department. Instead of ash trees they were asked to use *chir* trees, totally unsuitable for the purpose. The villagers faced utter injustice when the government allotted ash trees to the Symonds Co., a sporting goods firm. The UP Government tried to compromise with the people by allotting the DGSS ash trees on condition that the company would be allowed to take its quota. But the resistance made by the villagers forced the company to run away from Mandal without felling even a single tree. But the government went ahead with its plan of the yearly auction of forests despite the warning protests by the villagers.

A series of this kind of protests took place in the same period in many parts of India like Dehradun, Uttarkashi, etc., of Garhwal and Nainital of Kumaun, as a result of the decision of the State governments to manage the forests on commercial lines by selling the products to private companies and denying the original inhabitants of the region their rights for using forests. The *Appikko Chalewali Movement* that took place in the Western Ghats in the Uttara Kannada region of Karnataka in the 1980s (Sheth, 1997), a movement solely led by a group of women in the Asna village of Bastar district in the State of Madhya Pradesh in 1987 (Sundar, 1998), etc., are examples of organized protests against State governments.

The Narmada river valley project forms the most important case of study that pictures how the short-term, current development policies of the various State governments and the politics encompassing it can result in the pauperization of a large mass of people and in their continuing struggle for the right to live (Sangvi, 2002). The Narmada River Valley Project is a scheme envisaging the construction of 30 major dams, its tributaries, 135 medium and 3,000 minor dams on the Narmada, which has a history beginning from as early as the 1950s. It is an inter-State multi-purpose project comprising the four major States of Western India, viz., Madhya Pradesh, Gujarat, Maharashtra and Rajasthan, with a terminal major dam in Gujarat. The focus of popular opposition has been the Sardar Sarovar reservoir, the largest of the project's individual schemes, which will displace a large number of tribal families from their naturally lived environments and this called into question the policy behind it. This case witnessed the most popular ways of protesting by the affected people that could make an impact on the project at many stages of dam construction by temporarily halting the works at the dam site. *Narmada Dharmagrastha Samiti* (NDS), i.e., Narmada Displaced People's Organization, *Khedut Mazdoor Chetna Sangathan* (KMCS), i.e., Farmer Labourer Organization, an organization fighting from the early 1980s on the issues of land and other resource rights, livelihood and self-respect of the *adivasis, Narmada Asargrasta Sangharsha Samiti* (NASS), i.e., Narmada affected people's struggle society, *The Lok Adhikar Sangh,* (LAS), i.e., Association for People's Power, etc., are some of the organizations where the tribal people united for protesting against the State governments. These protests raised fundamental questions of their right to information, right to life, land and other natural resources and also the right to participate in the decision making processes. The victory of the oustees of other dams completed during the 1970s and the 1980s over the government of Madhya Pradesh was a successful assertion of their rights and became a milestone in the tribal resistance movements following the inception of the Narmada project.

The Tehri dam project on the river Bhagirathi in the Garhwal region of the Himalayas is another example where the government met with the local people's resistance. There are also protests that are still going on against the establishment of industries, impoverishing the natural inhabitants of the area, of their resources like water, air. etc. We can find a number of such resistance movements from around the world also, like the protest against the Bali project in Indonesia, the movement for the survival of the Ogoni people living on the plains of the Niger Delta, east of Port Harcourt, the capital city of the Rivers State in Africa, etc.

All these cases bring forth the fundamental questions of 'right' and 'justice' to live in their lived spaces; where there is violation of these rights; it became a life and death struggle for these people. Like any other human being, irrespective of the class, caste and other socio-economic and political positions, etc., lived space is of utmost importance to them because it is the determining factor of their survival and livelihood. The lived space is one on which they are dependent in ways more than one. It is like their habitat, though not necessarily natural, but that has become so by regular interaction. There is also the aspect of being part of a system of relations in which living beings interact with the environment and also among themselves, in that setting. Therefore, this habitat can rightly be called an eco-habitat in the sense of people both intervening and depending on it. The lived space, i.e., the eco-habitat is also a space around which the people weave an emotional bond. Their identity is something they come to attain in this process, which is affirmed, from time to time, by their lived experience in their eco-habitat. This adds value to eco-habitat as a culturally significant system. Now the eco-habitat of a people can be defined as a system, (i) in which people find themselves in a relation of co-existence with the rest of the environment, (ii) to which they are physically dependent on for their livelihood, and (iii) one that provides an enduring setting for their cultural identity. Therefore, when there is an external interference or a disturbance in the eco-habitat, it will have an impact on the life of the people.

The above-mentioned cases are examples where the eco-habitat of the tribal groups is either lost or facing threat of displacement. The interference detrimental to the eco-habitat of a people mostly comes from the State itself. Along with the State, different other human agencies authorized by the State or those who obtained legal sanctions from the State such as government officials, corporate bodies, political leaders, legislators, NGOs, members of communities, etc., also become parties in interfering with the eco-habitat of the tribal people. There is the presence of a 'power factor' that controls the relations between these different human agents in a specific manner. The case studies reveal that the powerful agents, i.e., those having access to wealth or having capacity for political lobbying, etc., try to influence the decision-making activity of the governments, which usually have an adverse effect on those who are the real claimants to the locally available natural resources in the existing socio-economic and political system. Such situations of social conflict becomes complex, when the number of dominant agents involved in it is too many and the relations between them take multifarious dimensions as determined by power. Thus the question (politics) of power often outgrows other concerns of vital importance, including the question of survival, for a larger group of people. Similarly, the approach to nature or environment is often influenced by the ways in which people are related among themselves politically and this will have a bearing on the tactics adopted for tackling the environmental (and also social) problems. That is, many environmental conflicts are driven by dominant power relations over the environment, which often benefit the few, threatening the survival of many.

It is the authority to intervene in an environmental setting that renders the human agency its political significance. When the impact of such activities negatively affects the eco-habitat of a local people and they recognize this and decide to resist the move to intervene in their eco-habitat, then their reaction also assumes political significance. In other words, the intervention into an eco-habitat by the external agencies and people's resistances when their eco-habitat is affected reveals that the political dimension is internal to the eco-habitat. It points to people's tacit recognition of the arrangement of a given eco-social life and an understanding that the power to continue or change it lies solely with the people. Human beings peopled in an eco-habitat are perceived to have an inalienable right to preserve, continue or change the eco-social life that it gives rise to. This is the 'political' in the eco-habitat and is reflected in the many environmental movements around the globe.

Not only clash of interets but also clash of values may lead to a situation in which eco-habitat problems become politically impregnated. By clash of interests, we mean different agents coming into conflict over the same resources. In conflict of values, different groups confront each other on the basis of the different values they attribute to living or non-living part of the environment. Sometimes conflicts emerge when values and interests of two groups are diametrically opposed. When their eco-habitat is disturbed, the affected people will find themselves in conflict with agents who initiated activities that led to such disturbances. In this context, we may speak of the dominant agents whose actions bring about negative changes in the eco-habitat of a people who are the affected agents. The dominant agents may or may not be aware of the consequences of their actions. If the disturbance of eco-habitat is an unintended consequence of the actions of these dominant actors, the responsibility for such actions does not lie with them. But, if they continue with their actions or do not find them incorrect they cannot be freed from the responsibility for their actions. If on the other hand, they were aware of the consequences of their actions that led to the eco-habitat imbalance, they are accountable for the outcome of the actions. Dominant agents are those, who while responsible for their actions, enjoy legal sanctions to act in the way they do, or have powerful political clout for carrying on with their activities. In both the senses, the situation is politically implicated: in the first sense, by being legally entrenched in the polity; in the second sense, by being politically influential. In the case of affected agents, if they fight for their right to protect their eco-habitat, then that struggle is political, in two senses: (i) If the situation arises because the laws of the land are silent or not clear about the legitimation of the actions of the dominant agents, then the struggle demands a change of situation through legislation; (ii) If the actions of the dominant agents are not legally protected, but get support from and through political lobbying, then the affected agents can try to pressurize the government by means of civil society movements to deliver justice to them.

Politics involves relations of power between agents with respect to their positions in the field where they act. When the question of authorization of resources arises in a field, the power relations between agents become strained. When such authorization of resources is legally entrenched, the political describes the arrangement of institutions pertaining to the field. Resolving to keep the existing arrangement or recognizing the need to change it both can arise from material interests of the agents as well as from their value orientation. In the context of eco-habitat, in most cases, the values the inhabitants attribute to the *status quo* of the eco-habitat and their interest in it are the same. The political question in this context is whether the affected agents have the authority to make changes in the arrangement or preserve the preferred one in their eco-habitat. It is simultaneously also a question of whether the dominant agents have the authority to change the arrangement in the eco-habitat to suit their own interest.

It can be seen that the concept of 'eco-habitat' is not of local dimension alone, but has global dimension, as well. In cases of global issues such as global warming or ozone layer depletion, all categories of human beings fall prey to its various consequences. Here the sustenance of the eco-social life of any human being, i.e., of both the powerful and the powerless group, in the planet is threatened. In the case of issues of global dimension, the governments seem to be acting responsibly as these are problematic for each and every citizen of the country. As a result, various countries through international negotiations try to make consensus regarding the different environmental problems that have a wider impact. But when dealing with the local issues, the tribal people are not at all treated as citizens of a country and their problems are not taken care of. Rather the government seems to protect the interests of those, who are in power at the cost of the lives of the poor people. In most of the cases referred above, the governments seem to be influenced by the powerful agencies like the industries, multinational companies, political parties, media, etc., thereby putting into trouble the lives of the weaker sections of the society. But the fact is that in a democratic institutional set-up, the government is bound to protect the interests of those who are the real sufferers, by ensuring justice to them. There should be generality in the approach of the government in the sense that it should represent the stakes of all groups of people who are the real beneficiaries.

This chapter argues in this broader context that the local movements must be considered as equal to or more significant than the global environmental movements on the basis that (i) the tribal people also are the citizens of a country like the 'citizens' in any global issues and should be taken care of by the governments; (ii) the local issues of survival and livelihood that usually emerge as the immediate effects of any development programme affect the people in a shorter period and hence need immediate solutions, compared to the global issues, which have long term effects through a gradual process; (iii) The solutions to the tribal issues like that of the global issues are the responsibility of each and every citizen and hence are to be addressed by the civil society.

NOTE

1.The forest department was established first in 1864 and the Indian Forest Act was passed in 1878, according to which the forests were divided into reserved, protected and village forests. But even in the lands, which were retained for the use of the communities, they were denied all rights for regulating the use of resources. (Guha, 1989, p. 40).

REFERENCES

Chatterjee, Partha (1994). *The Nation and its Fragments : Colonial and Postcolonial Histories,* Delhi: Oxford University Press.

Guha, Ranajit (1983). *Elementary Aspects of Peasant Insurgency in Colonial India,* Delhi: Oxford University Press.

Guha, Ramachandra (1989). *The Unquiet Woods : Ecological Change and Peasant Resistance in the Himalaya,* Delhi : Oxford University Press.

Sangvi, Sanjay (2002). *The River and Life : People's Struggle in the Narmada Valley,* Mumbai: Earth Care Books.

Sheth, Pravin (1997). *Environmentalism : Politics, Ecology and Development,* New Delhi: Rawat Publications.

Sundar, Nandini (1998). "Asna Women : Empowered or Merely Enlisted?", in Arne Kalland and Gerard Persoon (ed.), *Environmental Movements in Asia,* Surrey: Curzon Press.

17
Sacred Grove–Diversity in Forest Eco-System
A Study among the Kandhs of Kandhamal, Orissa

P.R. Sahoo and *P.M. Nanda*

Introduction

Studies reveal that since time immemorial, conservation of natural resources has been an integral aspect of many indigenous communities all over the world. It is also observed that traditional ethics of many tribal people, directly or indirectly protect forest patches by dedicating them to local deities. Such forest patches are referred to by these people as 'Sacred Grove'—a place of religious importance as well as a socio-cultural ground for observing various ceremonies. The Kandha of Orissa believes that success of any human attempt depends on the 'Supernatural Power which controls everything. The Sacred Grove for them is an abode of deities who remain inside the grove and care for the villagers, animals and plants. It is a holy religious place named as *Sala*. The grove (*Sala*) consists of a cluster of trees that exist in a circular shape and varies according to its nature and type. There are a number of groves around the village named after different deities. They consider their groves as mirroring their culture and have a family type of ties with the gods and goddesses who are believed to reside in the sacred grove. They believe that the trees provide them with all types of amenities like food, clothing, shelter, medicines, animal feeds, etc., and bring happiness and well-being for them. They also have a profound knowledge of their own environment, especially of their forest-based ecosystem. The Sacred Grove is one of a kind where trees of different nature are preserved. The knowledge about the utility of these has been passed on from generation to generation, mostly orally through stories, and cultural practices. The knowledge may be a simple life lesson, a tale with a moral ending or practical knowledge on cultivating crops, health care, and observation of ceremonies relating to their wellbeing; it is a tradition that they are expected to maintain.

In this context, an intensive field survey was conducted in different villages of Kandha tribe of Kandhamal, Orissa and data generated through questionnaires, focused group discussions, and documentation. This chapter emerges out of that larger study intended to understand the ethno-mindset relating to the Sacred Grove and document the plant resources having socio-cultural importance for the tribe, and the traditional and ethno-medicinal information that is usually preserved by the tribe in the Sacred Grove and also in their forest environment, besides other objectives of an action oriented nature, like cultivation and propagation of important medicinal plants, generating awareness among the people about the importance of different plants by extending Sacred Grove complex through afforestation and to get the indigenous knowledge relating to medicinal plants of the area patented.

Indigenous Knowledge

Indigenous knowledge (IK) is the local knowledge—knowledge that is unique to a given culture or society. IK contrasts with the international knowledge system generated by universities, research institutions and private firms. It is the basis for local level decision making in agriculture, health care, food preparation, education, natural resource management, and a host of other activities in rural communities (Warren, 1991)[1].

Traditional Knowledge (TK), indigenous knowledge (IK), and local knowledge generally refer to the matured long-standing traditions and practices of certain regional, indigenous, or local communities. Traditional knowledge also encompasses the wisdom, knowledge, and teachings of these communities. In many cases, traditional knowledge has been orally passed on for generations from person to person. Some forms of traditional knowledge are expressed through stories, legends, folklore, rituals, songs, and even laws. Other forms of traditional knowledge are often expressed through different means.

Indigenous knowledge can be defined as culturally and context-specific non-formal knowledge that is locally bound and indigenous to a specific area and orally transmitted, and that is generally undocumented, but at the same time dynamic, adaptive and holistic in nature and closely related to survival and subsistence of people (http://www.wikipedia.org).

The Protective Laws

Due to failure of pure legal protection in generating conservation, it has become imperative to search for alternative solution, based on local problems. Indigenous knowledge and practices need to be acknowledged to provide a lost advantage and edge to modern scientific conservation initiatives. Rules that pertain to resource use may ensure a sound religious system to find out a strong legitimacy that goes beyond the scientific arguments.

The Sacred Groves, irrespective of their nature and size, would find scope to blend an ecological function and a cultural justification of bio-diversity on the one hand and indigenous system on the other. Reforms are required in patent laws to ensure that small innovators and traditional knowledge can be protected, dispersed, and the disadvantaged innovators cannot be expected to win ever in the 'first to file' system. Special opportunity of 'first to invent system' should be made possible for them, community intellectual property rights to be enacted, each patent applicant to declare that knowledge and resources used for making claims have been obtained 'rightfully and lawfully', international registry to be created for grassroots innovations and traditional knowledge to reduce transaction costs of prospective entrepreneur and investors to meet innovators.

In Kandhamal, the most neglected area is the health sector, because modern facilities are yet to reach the interior areas. In such cases, the Sacred Grove fulfils most of the medicinal requirements of the people as they are devoid of health amenities and, therefore, compelled to depend on traditional medicinal practices. The Kandha preserved most of their essential medicinal plants in such groves for their own use and also looked after their sustainability in their own ecosystem. The plants they care for in the Sacred Groves are abundantly found in their local forest which is used by the medicine man for tribal health purposes. The tribal healers used to pick up most of their herbal medicines from these forest-based sacred groves.

The Structure of the Sacred Grove

The Sacred Grove in each village formed a unique shape along with the structure of the village. The Kandha people traditionally found their deities in such Sacred Groves. In the core of the sacred centres, there are 3-4 types of trees, the roots usually in a peculiar shape. The trees inside the Sacred Grove are more than 100 years old. The trees found in the central area are usually older than the peripheral ring of the outer circle. They spread over 2 to 6 acres of land. It is a cluster of trees of different nature. The groves are known according to its situation, direction, surrounding and availability of trees inside.

Generally the tree referred to as *Gramaseni* remains at the entrance of the village and keeps a watchful look out for the incoming outsiders and evil spirits and is like a Mother to all the villagers. *Giridevata* (*Sarupenu*), like a father guides the villagers and remains at the periphery of the village. Both of the *Giri Debata* and *Maa Pitabali*, according to the Kandhs, care for and guide the villagers as daughters and sons. They also believe that these trees/ deities protect them from evil spirits as well as from difficulties and mistakes. *Pitabali* is believed to ensure good health, good vegetation and well-being of the people, whereas *Jhakar kutti* (*Meri*) ensures good health, good harvest, rain and protection of plants and animal life.

Multi-purpose Plants in a Sacred Grove

Medicinal and Aromatic plants are extensively used throughout the world to cure human diseases from ancient period. In India about 47,000 plant species are distributed in the different vegetation zones, which is ranking eighth in the world biodiversity. Out of these plants about 8,000 species are known to be of medicinal importance. Around 2,500 plant species are used in the Indian System of Medicine such as Ayurveda, Unani, Siddha and Homoeopathy systems of medicines.

The trees in the Sacred Grove used for different purposes by the tribe are as follows:

***Sal tree** (Sargi)*

1. For constructing house
2. Use for firewood
3. Making furniture
4. Collecting flowers and *jhuna*
5. Preparing wood slab (*patta*).

***Banyan tree** (Bara)*

1. Deities are believed to reside on the tree and offer protection to devotees from evil eyes.
2. Use for fire wood
3. Fruits are taken by birds and animals

***Salap tree** (Salapa)*

1. Use for preparation of liquor
2. The trunk is used for making wood slab

***Blackberry tree** (Jammukoli)*

1. Blackberry seeds used for medicine
2. Wood for constructing House.

***Mango tree** (Maska)*

1. Leaves used in different religious occasions
2. From mango preparing sara, champeyat, ambula.
3. Wood is used for firewood and preparing wooden slab

***Jack fruit tree** (Panasa)*

1. Fruits are eaten
2. Firewood and construction of house
3. Use for furniture.
4. Leafs are used in different religious ceremonies.

Bamboo tree *(Baunsa)*

1. Use in construction of house.
2. Preparing baskets
3. Preparing household commodities
4. Leafs used as cattle feeds and medicine

Bhalia tree

1. Use as firewood
2. Prepare medicines from seeds
3. The seeds used in magical practices. Small branches used for making fences in paddy fields

Kurei tree

Use its flower for worship in different occasions

Kris

Use for wood and house construction

Harida, Bahada

For medicinal use

Sarpadali

Construction of house and use as firewood

Pendal, Pipal, Karanja

For medicinal use

The traditional knowledge about plants and animals of the tribal people has the potential for commercial application in industrial practices, medicine and health. This needs to be documented to preserve such indigenous knowledge from near extinction.

Conservation

The Sacred Groves in the forest are now at stake due to lack of proper preservation and protection. Most of the trees are caught indiscriminately by the forest mafias as well as by the poor inhabitants of the villages due to non-supervision of conservators. The rapid deforestation not only affects the forest ecosystem of the area but also destroyed the socio-cultural set-up of Kandha villages.

Conclusion

The Kandha tribal ethics centering round the vegetation and their deities *Pitabali, Meri, Gramaseni, Buru penu, Tari penu, Darni penu,* and *Saru penu* (*Parbata devata*) are connected to more than 300 species of various nature. There is an urgent need not only to protect the grove but also to revive and reinvent such traditional practices of natural conservation, and encourage people to protect, preserve and regenerate sacred groves in tribal areas to maintain bio-diversity and ecosystem. It is imperative that sacred groves be explored in terms of ecological and cultural perspectives in order to develop a strategy for revival of the tradition in conservation. Therefore, this chapter suggests that attempts should be made to explore the various socio-cultural and ecological dimensions of sacred groves maintained traditionally. Important medicinal plants are becoming rare and endangered due to over exploitation. In this context, there is a need to introduce new agro techniques, utilize waste lands for medicinal plants cultivation, and conserve and get patented traditional and ethno-medicinal plants which are our National Heritage. Sacred Groves of tribes should be preserved as Cultural Heritage sites.

It is also suggested here that the forest patches may be protected on a co-sharing basis by the villagers and Government and to ensure commercial viability and enrich tribal economy while encouraging the villagers to be involved in the process of conservation of forest-based ecosystem, eco-friendly forest parks need to be created to augment both forest revenue and village economy through nursery, training centers and tourist arcades attached to Sacred Groves, forming Sacred Grove Complexes. Such Complexes are to be well protected through government and nongovernmental partnerships and awareness creation among people on conservation of forests and their ecosystems.

NOTES

1. Warren, D.M. (1991). "Using Indigenous Knowledge in Agricultural Development"; World Bank Discussion Paper No. 127. Washington, D.C.: The World Bank.

18
A Comprehensive Environmental Conservation Project and Some Lessons

P.R.G. Mathur

Introduction

This chapter discusses a massive eco-restoration project, viz., the Attappady Wasteland Comprehensive Environment Conservation Project, initiated in an area that was heavily ecologically degraded and fragile with migrant peoples' infiltrations into the area and their practices, and the comprehensive conservation efforts that put in place a huge institutional network and infrastructural development. Highlighting the case study, this chapter also argues that if unchecked, resource conflicts and insecurities can interfere with such interventions with people's participation and involvement. The chapter also contends with the help of an illustration, that political interventions in decentralized governance structures and peoples' institutions will hamper the democratic functioning of such institutions and may call for strict corrective measures.

The Attappady Eco-restoration Project

The Attappady Wasteland Comprehensive Environment Conservation Project (AWCECOP) or the Attappady Eco-restoration Project in short, was a project for sustainable development funded by the Japan Bank for International

Cooperation (JBIC) with an investment of Rs. 219.321 crore (JapaneseYen 6,338 million) for restoring the ecosystem as well as the livelihood of the people of the Attappady valley which is well known as the most concentrated tribal area in Kerala.

The Attappady Block in Palakkad District of Kerala witnessed massive influx of plainsmen from Kerala and Tamil Nadu. These massive encroachments over forest and cultivated lands have been responsible for economic and cultural degradation manifest in the lives of the people of Attappady, especially that of the *adivasis*. These tribal people, belonging to three ethnic groups known as *Irula, Muduga* and *Kurumba,* have been reduced to a minority and pushed away to infertile hilltops due to large-scale influx of settlers from within Kerala as well as from the neighbouring States of Tamil Nadu and Karnataka. According to the Census of 2001, the population of Attappady was 67,672 of which 28,711 were the tribal. The tribal population had constituted a majority until 1961. But by 1971 they dwindled to a minority status in the valley.

The massive migration to Attappady also caused several ecological and environmental problems. The valley became highly fragile, landholdings became fragmented, soil and vegetation got degraded with unsustainable agriculture, and there was massive deforestation and uncontrolled grazing. One of the most important reasons for the ecological imbalance of Attappady was the widespread destruction of the natural plant species of the area. Overall degradation of the ecosystem was observed in all the three major categories of land in Attappady, namely, private agricultural holdings, tribal land, and the government land. It is in this context that the Attappady Eco-Restoration Project was taken up in 1996 with the objective of halting and mitigating these ecological and social degradations. The project aimed to reclaim the degraded wastelands in the Attappady region through sustainable developmental programmes implemented for land, water and biomass management and employment and income generation. All the programmes were evolved with informed and equitable participation of the beneficiaries.

Ecological restoration of degraded Wasteland in Attappady, development of replicable models to participate eco-restoration so as to prevent further degradation, and promotion of a sustainable livelihood for the local people (with special emphasis on tribal population) and in harmony with the resource base were the professed goals behind the Project. The Project commenced on 22nd April 2000 with the Attappady Hills Area Development Society (AHADS) as the implementing agency. Of the total cost of the project of Rs. 219.31 crore, there was a loan component of Rs. 176.9 crore and the share of the State Government was Rs. 42.41 crore. According to a report published by AHADS in March 2008, the project had spent a total amount of Rs. 137.52 crore until then, of which Rs. 117.30 crore had been utilized against the loan component and 20.22 crore as State share. Through the subsequent amendments to the loan agreement, the loan disbursement closure date was extended up to March 2010. Evaluating the programme as a success that can be replicated, Government of Kerala has decided to replicate it in Wayanad, another tribal concentrated area with unique ecological features. It is in this context that the participatory process of eco-restoration by the three indigenous communities together with the other settler population in the region is being examined. More than 5,000 people, as office-bearers, were directly involved in the implementation of the Project activities through 597 People's Institutions (PIs), of which 63 per cent were women and 85 per cent tribal people. The very fact that these communities were implementing Project activities worth millions of rupees speaks volumes about the goal of the Project to involve and empower the indigenous people in the eco-restoration efforts.

Institutional Arrangements for People's Planning for Watershed Development and Participatory Implementation

Three distinct but inter-linked levels of planning and management were adopted for the implementation of the Attappady Eco-Restoration Project :

1. Perspective planning at the project level through AHADS;
2. At the Watershed level through the Development Units (DUs); and
3. At the micro-watershed level through the User Associations (UAs).

The Micro-watershed level project planning and implementation was further differentiated into:

(a) At the level of activity locations through Local Action Groups (LAGs);
(*b*) At the Adivasi (tribal) hamlets through *Ooru Vikasana Samithis* (OVSs); and
(c) Interventions for forest conservation and afforestation through Joint Forest Management Committees (JFMCs).

The concept of watershed development recognizes water as the most fundamental, vital, natural unit of living systems on earth where its interaction with soil sustains life. Accordingly, the entire project area was scientifically mapped out as 15 inter-linked macro-watershed with natural boundaries. Two major rivers—Bhavani and Shiruvani—drain the project area. The Bhavani basin has three distinct watersheds and Shiruvani has two. These 5 watersheds were segmented into 15 macro-watersheds. Corresponding to these15 watersheds, fifteen development units were conceived as the middle level project management structure. These development units were further divided into 146 micro-watersheds.

User Associations

User Associations (UAs) were organized in the micro-watersheds to coordinate the implementation of micro-action plans. Beneficiaries of a micro watershed formed the User Associations, which executed all works pertaining to the micro-watershed. User Associations, registered under the Societies Act 1860, were formed in 93 micro watersheds out of the 146 micro-watersheds in the Project area. The Executive Committee of User Associations consisted of 9 members, of which a minimum of 5 woman members and 4 tribal members were mandatory.

Local Action Groups

As the micro-watersheds too were not separate entities but formed a mutually dependent and enriching system, a correspondingly interrelated and decentralized structure was designed for the project implementation location. Local action groups (LAGs) were organized as a participatory body of all the beneficiaries/users within a geographical unit of 50-100 ha in a micro-watershed.

Ooru Vikasana Samithis (OVS)

Recognizing the uniqueness of tribal societies such as their traditional kinship and community organization and to give special emphasis on ensuring their participatory development, exclusive *Ooru Vikasana Samithis* (OVS) (hamlet development societies) were formed in each tribal hamlet (*Ooru*) within the project area. Some 166 Hamlet Development Committees, exclusively tribal committees, were set up to take up development works in and around the tribal settlements. The Executive Committee of an *Ooru Vikasana Samithi* consisted of 13 members of which a minimum of 6 had to be women. These 166 OVS were organized covering the 187 tribal hamlets (in some cases, by combining closely situated hamlets).

A 10-point Hamlet Development Package was formulated for implementation by the OVSs. These packages provided for soil and water conservation activities, planting fruit trees, construction of pathways to the hamlets and common cattle sheds, nutritional foodgrain cultivation, propagation of smokeless *chulahs* and energy efficient hot boxes, establishment of biogas units and various income generation activities.

Joint Forest Management Committees (JFMC)

54 JFMCs were set up for taking up afforestation and protection works in the State forestlands. These were registered with the Divisional Forest Officer, Mannarghat. The Executive Committee of a Joint Forest Management Committee consisted of 9 members of which a minimum of 3 women were essential and priority had to be given to SC/ST members. Local forest officer, an *ex-officio* member, acted as Secretary to the committee.

Thaikula Sangam (TKS) (Mother's Group)

Total 111 Mothers' Groups came into being in Attappady. These were groups of tribal women who worked collectively as a social corrective force to fight against social evils like alcoholism, drugs etc. Registered as a Society, each Thaikula Sangam had an executive committee of 12 members.

Income Generation Activity Groups (IGA)

As part of sustainable livelihood options, Income Generation Activity Groups were also constituted by User Associations and *Ooru Vikasana Samithis*. The Income Generation Activity Groups aimed at promotion of saving habits, group cohesiveness, and entrepreneurship development for sustainable income and arresting of further degradation of Attappady. These groups were set up to take up sustainable income generation activities through promotion of micro-credit among members in the project area. They functioned as self-help groups and were integrated into the State-wide poverty alleviation programme–*Kudumbashree*. In all, 219 groups such groups came into being, each group consisting of 12 to 15 members, and with a total membership of 2,875. Promoting thrift collection, micro crediting, bank linkages and establishing sustainable micro enterprises were the activities implemented by these groups to ensure income to the poor even after the project support was withdrawn. Of these, 35 were bank linked groups, 128 were into micro enterprises, and they had developed group savings worth Rs. 16.92 lakh. Their total internal loan was Rs. 8.26 lakh and the external loan was Rs. 2.15 lakh.

Thus, for the eco-restoration of Attappady, a massive institutional structure was put in place with stakeholder participation undertaking various activities for a comprehensive development of Attappady. In a total of 644 People's Institutions, there were 7,163 Executive Members of which, Woman Executive Members were 4,741 (66%) and Tribal Executive Members, 6,088 (85%).

Special Focus on Tribal Development, but non-tribals benefited more

The Project implementation was planned in such a way that about 60 per cent of the project benefits would directly reach the tribal people. In Attappady, 41 per cent of the total populations are tribals. The activities such as Total Hamlet Development Programme, Environmental Literacy Programme and the construction of Community Resource Centre were exclusively for the development of tribal population. These programmes, along with construction of access roads to hamlets and other hamlet development programmes were implemented by the OVS themselves. The Project had the goal of working intensively for the empowerment of tribal people so that their informed participation in the development programme would be ensured.

Total Hamlet Development Programme

It is a programme aimed to address total habitat development plan for tribal areas, envisaged comprehensive development by incorporating the social, cultural and individual needs through participation in planning, execution, implementation and usage, ensuring transparent and tribal empowerment practices.

Housing and landscaping, accessibility, drinking water supply, environmental hygiene and sanitation, sustainable farming systems and livelihood generation were the major components of the programme. These were being implemented in 24 hamlets through 13 programmes including housing for tribals.

In the afforestation programme, however, it was found that instead of afforestation of wastelands owned by the tribals, seedlings were planted in the developed gardens of non-tribals. This happened in Northommala and Guddayur. The encroachers of the forest and tribal land benefited more from the eco-restoration programme compared to the tribal people without any cost-sharing.

Conflict over Resources–A Case Study

The conflict between the JFMC Mele Chootara and Umathampadi JFMC over supplying water from the former (Aun-Penn) forest top slip to the latter is a very insightful case study and, therefore, worth elaborating upon.

The officials from AHADS visited Chootara and held discussions with Rassi, the President and other members of JFMC. The Field Officer (Forestry) also arranged a meeting of the General Body of JFMC of Chootara, which was attended by 60 persons. The meeting held on 16th April, 2000, and passed a resolution for bringing water from Chootara top slip to Umathampadi. It was also resolved to construct a water tank at Malliswaram Koil Umathampadi to supply drinking water to the tribal and non-tribal families of Chootara and Umathampadi. The President of JFMC was authorized to take necessary steps for constructing the gravity of drinking water system.

In July 2003, an agreement was executed between the JFMC, Chootara and UA 4/3 Umathampadi for the construction of a gravity irrigation system for drinking water. The proposal was for a scheme costing Rs. 3 lakh, to bring water through pipes from the top slip of the Chootara (Aun-Penn) to Umathampadi. Rassi, the President of JFMC, Chootara, W/o. Kalimuthu, a non-tribal who belongs to the Boyan caste (a Scheduled Caste), signed the agreement between the JFMC Chootara and UA 4/3 Umathampadi. The gravity irrigation work was started and an amount of Rs. 1.5 lakh was paid as advance out of the sum of Rs. 2.66 lakh already set apart. A tank with a capacity of 10,000 litres was constructed at Umathampadi. The UA 4/3 purchased pipes worth Rs. 1.5 lakh. The tribal people of Chootara feared that if water were given to Umathampadi, the Chootara Thodu (Udragar) would dry up. The Irulas of Thazhe Chootara and Mele Chootara challenged the agreement reached between the JFMC and UA 4/3 for installing gravity water system from the top hill of Chootara. It is alleged that one day two Irula men, Kokkala *alias* Nanchan and Chathan, were returning from work they attacked Rassi and her husband, Kalimuthu, with stones and fractured Kalimuthu's collarbone. The couple was admitted to Agali Hospital by the Field Officer (Forestry), AHADS and later removed to the Palakkad District Hospital for treatment. They were hospitalized for about a week and the JFMC had to spent Rs. 2,500 for the treatment.

According to the Field Officer (Forestry), Rassi had once lodged a complaint before the Range Officer, Pudur, against the *ganja* cultivation practiced by the tribals of Mele Chootara, thus incurring the wrath of these people. Similarly, Venugopal, Deputy Ranger of AHADS observed that there was sufficient water in the upstream of Chootara.

Rassi, as President of the JFMC took the decision in supplying water through gravity system against the interests of the tribal community. She had to leave Mele Chootara hamlet and take refuge at Mele Umathampadi S.C. Colony since the tribals of both Mele Chootara and Thazhe Chootara threatened her life.

The tribals of Thazhe Chootara alleged that Rassi stabbed Nanchan, (Kokkala) brother of Ponnan; the President of OVS with her sickle and his knee sustained a five-inch deep injury, when he and his friend Chathan went to her house and questioned her authority in supplying water to Umathampadi. Nanchan has become invalid and his wife Mari has to look after him and his four daughters of whom two are students at Sholayur School. The tribals alleged that Rassi did not consult the General Body before taking the decision to supply water through gravity system. All villagers opposed her decision and the General Body of JFMC dismissed her from the office of President. Nanchan also refuted the charge that he attacked Rassi with stones.

Ponnan, President of OVS, Thazhe Chootara and Secretary, U.A. Olavankara said that his brother Nanchan was attacked by Rassi with sickle and he had to spend Rs. 20,000 for the treatment of his brother at Thadakam Hospital. Malli (65) an Irula woman of Thazhe Chootara said that Rassi would be beaten to death and her body would be carried to the Police Station.

Rassi's version was that Nanchan *alias* Kokkala came to her house at 8 p.m. after getting intoxicated. Her husband Kalimuthu was then taking rest in the outer verandah of their house after his return from work at Manthe Chalan Gowder's house. On that day she had gone to the office of AHADS, Bhootavashe, and had withdrawn Rs. 5,600 to pay wages to the watchmen attached to the JFMC. She was preparing dinner when Nanchan and Chathan came and attacked her with stones. She was first admitted to Agali hospital and afterwards removed to the district hospital.

The JFMC had earlier implemented a scheme for the supply of drinking water to Mele Chootara by spending Rs. 1 lakh through gravity system from the Chootara top slips, but the system had collapsed after six months. The tribal people had to take water from a neighbouring stream. On account of substandard construction and financial irregularity, the President of the JFMC earned the wrath of the tribal beneficiaries. This appears to be the backdrop of the conflict over resources and attack. The steps taken by the General Body of the Chootara JFMC in asserting its power by dismissing the President from the office and leading to her banishment from the hamlet, forcing her to take refuge in a neighbouring SC colony might have been extreme, but it brings to fore the fact that internal conflicts are often unleashed when resource constraints are acutely felt.

Corruption and Political Interference—An Illustration

It is undisputable that corruption interferes with the democratic functioning of peoples' institutions. We find that the General Body

Index